Basic Electricity

SECOND EDITION

A Self-Teaching Guide

CHARLES W. RYAN

JOHN WILEY & SONS, INC.
New York • Chichester • Brisbane • Toronto • Singapore

For my son, Dan Eric Ryan

The tables of Natural Trigonometric Functions in Appendix II are reproduced by permission from Essentials of Mathematics, 3rd ed., by Russell V.
Person (New York: John Wiley & Sons, 1973).

Library of Congress Cataloging-in-Publication Data

Ryan, Charles William, 1929—
    Basic electricity.

    (Wiley self-teaching guides)
    Bibliography: p.
    Includes index.
    1. Electricity—Programmed instruction. I. Title.
II. Series.
QC523.R9 1986   537'.07'7   86-11114
ISBN 0-471-85085-3

Printed in the United States of America

86 87 10 9 8 7 6 5 4 3 2

Preface to the Second Edition

During the ten years this book has been in print, the technology of electronics has been advanced in important ways; but the basic principles of electricity have changed not at all. And the method of instruction used in the Self–Teaching Guides has stood the test of time.

When you complete each "frame" of instruction, you know instantly whether you have absorbed that unit of material. A self–test at the end of each chapter allows you to evaluate your understanding of the major points covered in the chapter as a whole. An end-of-course examination has been added in this second edition so that you may test your overall comprehension of basic electricity.

The self–tests have been revised, and minor changes have been made to the text. A cross–reference list of standard texts on electricity has been brought up to date, and a short afterword has been added. The book as a whole remains consistent with the presentation in the first edition. A very practical and often–quoted rule applies here: "If it isn't broken, don't fix it."

Charles W. Ryan

Spokane, Washington
April 1986

Acknowledgments

I am grateful to the following individuals and one government agency for their assistance and cooperation in preparing this book:

The Bureau of Naval Personnel, United States Navy, for the use of many illustrations that originally appeared in Rate Training Manual NavPers 10086-B, Basic Electricity.

Louis Wertman, Chairperson, Electromechanical Technology, New York City Community College, for reviewing the manuscript and offering many helpful suggestions.

Stewart R. Powell, Associate Professor of Electronics, Orange Coast College, for reviewing the manuscript and offering many helpful suggestions.

Richard Rotter, Monterey, California, for assistance in the preparation of illustrations.

Irene F. Brownstone, John Wiley & Sons, Inc., for helpful guidance in programming the material.

John H. DeChenne II, whose technical contribution to the second edition is greatly appreciated.

My wife, Elaina Jean Ryan, for patience, understanding, and moral support during the writing of this book.

Contents

Cross-Reference Chart

The following table identifies the chapters in this book with corresponding material in a number of widely used texts. You may find it useful in looking for different treatments or additional material on specific topics or in relating work in this book to a course that uses one of the texts listed here. Following is a bibliography of the texts listed in the chart.

Blitzer, Richard, Basic Electricity for Electronics (New York: John Wiley & Sons, 1974).

Boylestad, Robert L., Introductory Circuit Analysis, 4th ed. (Columbus, Ohio: Charles E. Merrill, 1982).

Grob, Bernard, Basic Electronics, 5th ed. (New York: McGraw-Hill, 1984).

Jackson, Herbert W., Introduction to Electric Circuits, 5th ed. (Englewood Cliffs, New Jersey: Prentice-Hall, 1981).

Leach, Donald P., Basic Electric Circuits, 3rd ed. (New York: John Wiley & Sons, 1984).

Zbar, Paul B., Basic Electricity: A Text-Lab Manual, 5th ed. (New York: McGraw-Hill, 1982).

Chapter in This Book	Blitzer	Boylestad	Grob	Jackson	Leach	Zbar
1 What is Electricity?	1	2	1, 9	1, 3	1, 2	9
2 Understanding Voltage, Current, and Resistance	2	2, 3	1, 11	2-4	2-5	7-10
3 The Simple Electric Circuit	2, 3	4, 5	1, 2	4-6	2-6	11, 12
4 Series and Parallel Circuits	3, 4	5	3, 4	6	6	13-20
5 Direct-Current Compound Circuits	5	6	5	7	6	14, 18, 21, 22
6 Magnetism and Electromagnets	9	10	12-14	10	11	28-30
7 Introduction to Alternating-Current Electricity	11	13	15	14	14	30
8 Inductance	11, 12	11	16, 22	12, 13	12	44-46
9 Capacitance	13	9	19, 22	8, 9	13	48, 49, 51, 52
10 Inductive and Capacitive Reactance	12, 13	13	17, 20	15	15	44, 50
11 Alternating-Current Circuit Theory	11-15	13-20	23, 25	16-18	16-18	54-66

Introduction

It would be hard to imagine our modern world without electricity. Electricity pervades our lives at every turn, and its applications are almost endless. It has made possible an almost infinite variety of tools, most of which we take for granted, from the simple flashlight to incredibly complex computers and the telemetry systems that have already allowed us, through far-ranging space probes, to examine the outer planets of the solar system.

This book is not primarily concerned with applications but with the basic principles of electricity underlying all the applications. It will help you gain an understanding of the laws governing voltage, current, resistance, and power, as well as the somewhat more complex concepts of alternating current, which are the basis for electronics.

A knowledge of the simple equations of basic algebra is necessary to the study of electricity. The "short course" in Appendix I teaches all you need to know in this book, if you do not know (or have forgotten) how to solve an equation for one unknown. You might wish to test your knowledge of simple equations by taking the Self-Test at the end of Appendix I.

In later chapters, you often will work with very large or very small numbers. The use of powers of 10 is not required for these problems, but it could make your calculations easier. An optional brief review of the laws of exponents, including the powers of 10, is provided in Appendix III.

The material in this book is in the form of programmed instruction, a technique that helps you learn more effectively by getting you directly involved with the information as you read it. The material is presented in numbered segments called frames. Each frame presents some new information and poses a question or problem. Before you start reading, have a pencil and an index card (or folded piece of paper) ready. You will also need some extra paper for your work in frames that require you to solve problems.

When you reach each numbered frame, use the card to cover the answer below the dashed line. Read the instruction and question, do what is asked for in the frame, and then compare your answer with the one given in the book. If you are correct, go ahead to the next frame. If you made a mistake, be sure you understand why and correct the error before you go on. You will learn more effectively if you actually write out all your answers before looking at those in the book and if you correct any mistakes as you go.

Work at your own pace. You may be able to move rapidly through some sections, but only slowly and thoughtfully through others. Breaks are provided in most chapters, so be sure you complete a section or chapter at one time. Do not break in the middle of a section. You may wish to do only one chapter at a time with a few hours or a day in between chapters. Because of the continual review in answering questions, you will not forget one chapter if you wait awhile before beginning the next.

Each chapter ends with a Self-Test to help you evaluate your mastery of the concepts presented. Before going on to the next chapter, take the Self-Test, compare your answers, and review the indicated frames for any problems you missed. If you miss many items, review the whole chapter. A comprehensive final exam can be found at the end of this book.

The Cross-Reference chart on page vii correlates BASIC ELECTRICITY with some popular textbooks. You will find this chart useful if you are in a course using one of these books, or if you want to learn more about some of the topics in this Self-Teaching Guide.

If you work through the book systematically, you will be surprised at how easily you acquire an understanding of the fundamentals of electricity.

CHAPTER ONE

What is Electricity?

The word "electricity" comes from the Greek word <u>electron</u> ($\epsilon\lambda\epsilon\kappa\tau\rho o\upsilon$), which means "amber." Amber is a translucent yellowish mineral made of fossilized resin. The ancient Greeks used the words "electric force" in referring to the mysterious forces of attraction and repulsion exhibited by amber when it was rubbed with a cloth. They did not understand the nature of this force and could not answer the question, "What is electricity?" Today, we still cannot answer the question, although the success with which we have used electricity is obvious everywhere.

Although we don't really know what electricity is, we have made tremendous strides in harnessing and using it. Elaborate theories concerning the nature and behavior of electricity have been advanced, and they have gained wide acceptance because of their apparent truth—and because they work.

Scientists have found that electricity behaves in a consistent and predictable manner in given situations or when subjected to given conditions. Scientists, such as Faraday, Ohm, Lenz, and Kirchhoff, have described the predictable characteristics of electricity and electric current in the form of certain rules, or "laws." Thus, though electricity itself has never been clearly defined, its predictable nature and ease of use have made it one of the most common power sources in modern times.

By learning the rules, or laws, about the behavior of electricity, you can "learn" electricity without ever having determined its fundamental identity.

When you have finished this chapter, you will be able to:

- describe free electrons;

- describe conductors and insulators in terms of the movement of free electrons, giving examples of each;

- relate the action of free electrons to the phenomenon of static electricity;

- describe positive and negative charges;

- state the law of attraction and repulsion of charged bodies;

- explain Coulomb's Law of Charges; and

- describe the electric field associated with charged bodies.

Free Electrons

1. The classical approach to the study of basic electricity is to begin with the "electron theory." This encompasses the nature of matter and a fairly thorough discussion of molecules and atoms. Such an approach provides a good background for the essential point: Electric current depends on the movement of free electrons. In this book, the details of electron theory, such as atomic weights and numbers, are omitted so that we may move quickly to the points you really need to know for the study of electricity.

   All matter is made of molecules, or combinations of atoms, that are bound together to produce a given substance, such as water or salt or glass. If you could keep dividing water, for example, into smaller and smaller drops, you would eventually arrive at the smallest particle that was still water. That particle is a molecule, which is defined as the smallest bit of a substance that retains the characteristics of that substance.

   The molecule of water is known in chemical notation as H_2O. That means the molecule is actually made up of two atoms of the element hydrogen (H) and one atom of the element oxygen (O). These atoms, themselves, are not water but the separate elements of which the molecule of water is composed.

   What is the relationship between atoms and molecules? _____

   _____

   - - - - - - - - - -

   Molecules are made up of atoms, which are bound together to produce a given substance.

2. The ancient Greeks had conceived the idea of the atom, at least in theory. In fact, atom is a Greek word that means, roughly, "not able to be divided." Today we know that the atom is composed of even smaller particles. The most important of these are the proton, the electron, and the neutron. These particles differ in weight (the proton is much heavier than the electron) and charge. The weights of the particles need not concern you, but the charge is extremely important in electricity. Perhaps you have noticed that the terminals of the battery in your car are marked with the symbols "+" and "−" or even with the abbreviations POS (positive) and NEG (negative). Many batteries used in flashlights, small electronic calculators, and other devices have similar markings. The concepts of "positiveness" and "negativeness" will become clear later. For the moment, you only need to know that the proton has a positive (+) charge, the electron has a negative (−) charge, and the neutron is neutral, which means that its positive and negative charges are in balance. Practically speaking, we say that the neutron has no charge.

   The following drawing shows the relationship of visible matter to molecules, atoms, and smaller particles: electrons, protons, and neutrons.

The drawing shows only the electrons and protons in the atoms, but every atom except hydrogen also contains neutrons.

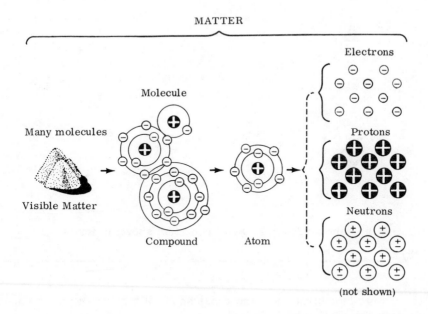

List the three basic particles that make up the atom, and state the charge (negative, positive, or neutral) on each.

Particle	Charge
_____	_____
_____	_____
_____	_____

- - - - - - - - - -

The proton is positive (+), the electron is negative (−), and the neutron is neutral.

3. As the drawing in frame 2 indicates, the atom has a nucleus (or core) that is positive because it contains only protons and neutrons. Electrons are in orbit about the nucleus, in much the same way as the earth orbits around the sun. A stable atom has the same number of electrons in orbit as it has protons in the nucleus. The nucleus always (with one exception) contains neutrons, too, but we need not consider them for our purposes because they are always neutral. Since the negative charge of the electrons is balanced by the positive charge of the protons, the atom is electrically neutral. The following drawing shows a hydrogen atom, the only one that has no neutron, so its nucleus is a single proton.

Label the electron and the proton.

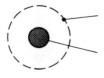

- - - - - - - - - -

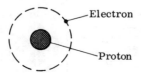

4. What is the net charge on the hydrogen atom shown in frame 3? _____

   Why? _____

- - - - - - - - -

The charge is neutral, because it has an equal number of protons and electrons (one of each).

5. Other atoms have more protons in the nucleus and more electrons in orbit. In fact, each atom has a different number of electrons and protons. In all cases, however, the electrons move around the nucleus of the atom in various orbits. Such electrons are not free: They are locked into the atom because they are attracted by the nucleus. They do not fall into the nucleus (as the earth does not fall into the sun) because their movement in orbit provides an equalizing centrifugal force.

   Free electrons are produced when some force disturbs the stable relationship of electrons and protons in an atom. This force, which "knocks" electrons out of orbit, can be produced in a number of ways, such as: by moving a conductor through a magnetic field; by friction, as when a glass rod is rubbed with silk; or by chemical action, as in a battery. (The six principal methods of producing this force, called voltage, are described in Chapter 2.) The force "frees" the electrons from their atoms; these electrons are called <u>free electrons</u>. When an electric force is applied to a material such as copper wire, electrons in the outer orbits of the copper atoms are forced out of orbit and impelled along the wire. The electrons

   that have been forced out of orbit are called _____.

- - - - - - - - -

free electrons

6. The movement of free electrons along a wire is what we call <u>electric current</u>. It cannot exist where there are no free electrons. What are free electrons? _____

   - - - - - - - - - -

   Electrons that have been forced out of their atomic orbits (or any similar wording that makes this point).

7. Explain in your own words the flow of electric current in a copper wire when an electric force is applied to the wire. _____

   _____

   _____

   - - - - - - - - - -

   Something like: When an electric force is applied to a copper wire, free electrons are displaced from the copper atoms and move along the wire, producing electric current.

Conductors and Insulators

8. Electric current moves easily through some materials but with greater difficulty through others. Let us see how the action of free electrons is related to current flow through these materials. Substances that permit the movement of a large number of free electrons are called <u>conductors</u>. Copper wire is considered a good conductor because it has many free electrons when an electric force is applied to it. Electrical energy is transferred through a conductor by means of the movement of free electrons that migrate from atom to atom inside the conductor. Each electron moves the very short distance to the neighboring atom, where it replaces one or more electrons by forcing them out of their orbits. The displaced electrons repeat the process in other nearby atoms until the movement is transmitted throughout the entire length of the conductor.

   The movement of each electron takes a very small amount of time, but the electrical <u>impulse</u> is transmitted through the conductor at the speed of light, or 186,000 miles per second. To see how this is possible, imagine a line of billiard balls that almost, but not quite, touch. When the ball at one end is struck by the cue ball, the ball at the other end is knocked away from the line almost instantly. The force travels through the line of billiard balls much more rapidly than each individual ball moves. This is basically how the electrical impulse travels. (Keep this in mind later in the book, when events in an electrical circuit seem to occur simultaneously.)

   Does electric current more closely resemble the actual movement of free electrons or the impulses transmitted as the electrons bounce against one another? _____

   - - - - - - - - - -

the impulses

9. Silver, copper, and aluminum all have many free electrons (the electrons
   are said to be "loosely bound") and are thus good conductors. Copper is
   not as good a conductor as silver, but it is the most commonly used material for electrical wiring because it is a relatively good conductor and
   is much less expensive than silver. Here are six metals listed in the order of the ease with which electrons are displaced from the atoms:

   > silver
   > copper
   > aluminum
   > zinc
   > brass
   > iron

   If we say that silver is a better conductor than iron, what do we mean?

   _____

   - - - - - - - - - -

   Electrons are displaced from silver atoms more easily than from iron
   atoms.

10. Some substances, such as rubber, glass, and dry wood, have very few
    free electrons; the electrons are said to be "tightly bound." Such sub-
    stances are <u>poor</u> conductors and are usually called <u>insulators</u>. Circle
    the material that is the best <u>insulator</u> among those named:

    > silver   glass   zinc   brass

    - - - - - - - - -

    glass

11. A good conductor, then, has many free electrons, while a good insulator
    has few free electrons. Dry air, glass, mica, rubber, asbestos, and
    bakelite are all good insulators. If we say that dry air is a better insu-

    lator than bakelite, what do we mean? _____

    _____

    - - - - - - - - - -

    Dry air has fewer free electrons than bakelite.

12. Write the materials listed below in the appropriate columns on the follow-
    ing page as either conductors or insulators. (Don't worry about the exact
    order.)

    > glass   silver   rubber   dry air   copper   brass   mica

Conductors   Insulators

- - - - - - - - -

Conductors: silver, copper, brass
Insulators: glass, rubber, dry air, mica

13. Define a good conductor. _____

- - - - - - - - -

A material that has many free electrons.

14. Define a good insulator. _____

- - - - - - - - -

A material that has few free electrons.

15. Name three materials that are good conductors. _____

_____

- - - - - - - - -

These have been mentioned in this chapter: silver, copper, aluminum,
zinc, brass, and iron.

16. Name three materials that are good insulators. _____

_____

- - - - - - - - -

These have been mentioned in this chapter: dry air, glass, mica, rubber,
asbestos, and bakelite.

Static Electricity

17. Static electricity is found in nature, so we shall examine this phenomenon
before we study "man-made" electricity. One of the fundamental laws of
electricity must be clearly understood to understand static electricity:
Like charges repel each other and unlike charges attract each other.
    A positively charged particle and a negatively charged particle will tend
to move toward one another. This is true even in a single atom. What
kind of particle will a proton attract, and why? Write your answer in the
blank on the next page.

_____

_____

- - - - - - - - - -

An electron; because the proton is positive, while the electron is
negative.

18. The electrons do not actually move toward the protons in the nucleus of
an atom. The force of attraction is resisted by centrifugal force. If you
have ever swung a bucket of water in a circle, you have seen an example
of centrifugal force. The force of gravity was not able to pull the water
out of the bucket, because centrifugal force opposed gravity and kept the
water in. In the same way, electrons are kept in orbit around the nucleus
by centrifugal force, which resists the pull of the protons in the nucleus.
Why is there a force of attraction between the protons in the nucleus and

the electrons in orbit? _____

_____

- - - - - - - - - -

Because protons (positive) and electrons (negative) are unlike charges,
and unlike charges attract each other (or similar wording).

19. Will a proton attract or repel another proton? _____ Why?

_____

- - - - - - - - - -

Repel; because all protons are positive and like charges repel each other.

20. Two particles can have unlike charges even if neither is positive (or if
neither is negative). A neutron is neutral; that is, it is neither negative
nor positive. A proton is more positive than a neutron, so the two particles have unlike charges. Do the electron and the neutron have like or

unlike charges? _____ Why? _____

_____

- - - - - - - - - -

Unlike; because the electron is more negative than the neutron.

21. Two neutrons will neither attract nor repel each other, because they are
neutral; that is, they have no charge. For each pair of particles listed
on the following page, state whether the particles will attract or repel
each other.

(1) proton and electron _____

(2) proton and neutron _____

(3) electron and neutron _____

(4) electron and electron _____

(5) proton and proton _____

- - - - - - - - - -

(1) attract; (2) attract; (3) attract; (4) repel; (5) repel

22. State in your own words the law of attraction and repulsion. _____

_____

- - - - - - - - - -

Unlike charges attract each other and like charges repel each other.

23. Let us begin our study of static electricity with an experiment you can
try right now. Tear up some paper into small bits and place them on a
table or other hard, nonconducting surface. Now run a comb rapidly
through your hair a few times, then move the comb near the bits of paper.

What happened? _____

_____

- - - - - - - - - -

The bits of paper were drawn to the comb.   (At least, that's what
should have happened!)

24. The experiment you have just made is a demonstration of <u>static electricity</u>.
When two bodies of matter have unequal charges and are near one another,
an electric force (the force of attraction or repulsion) is exerted between
them. But, because they are not in contact—or are not connected by a
good conductor—their charges cannot equalize. When such an electric
force exists, and current cannot flow, it is called <u>static electricity</u>.
"Static" means "not moving." The electric force that exists under these
conditions is also called an <u>electrostatic</u> force. What two conditions are

necessary for static electricity to exist? _____

_____

- - - - - - - - - -

Two bodies of unequal charges must be brought near one another, and
current must not be able to flow between them.

25. Each atom, in its natural—or neutral—state, has the proper number of electrons in orbit about its nucleus. That is, it has the number of electrons that helps to give the element its identity. Thus, the whole body of matter composed of neutral atoms will also be electrically neutral. Matter in this neutral state is said to have no charge and it will neither attract nor repel other neutral matter in its vicinity.

The atom of each element, in the neutral state, has a different number of electrons in orbit. Hydrogen has one, helium has two, etc. A model of an aluminum atom is shown below. It has 13 electrons in orbit balanced by 13 protons (P) in the nucleus. The nucleus also contains neutrons, but they need not concern us because they are electrically neutral.

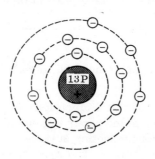

The aluminum atom shown is electrically neutral. Why? _____

_____

- - - - - - - - - -

Because it has the same number of electrons and protons.

26. But the three electrons in the outer orbit are easily displaced. If one (or more) of the electrons is knocked out of orbit, what is the charge on the atom? _____ Why? _____

_____

- - - - - - - - - -

Positive; because there are more protons (which are positive) than electrons.

27. Most substances, such as the hair, paper, and comb used in your experiment, are compounds rather than elements. That is, they are composed of atoms of various elements bound together in molecules of the substance. Their electrons are not easily displaced by an electric force. (You might have guessed that hair, paper, and combs are all relatively good insulators.) But the electrons can be displaced by friction, and this is what happened when you combed your hair vigorously in the experiment.

At the beginning of the experiment, your hair, the comb, and the paper were all electrically neutral. When you combed your hair, friction displaced electrons from your hair, and they were collected on the comb. At that point, what was the charge on the comb? _____ Why?

_____

- - - - - - - - - -

Negative; because the comb had an excess of electrons.

28. After the electrons were accumulated on the comb, bits of the neutral paper were attracted to the comb. Why? _____

_____

- - - - - - - - - - -

The comb and paper had unlike charges, and unlike charges attract.

29. One of the easiest ways to create a static charge is by friction. Two pieces of matter are rubbed together, and electrons are "wiped" off one and deposited on the other. The materials used can't be good conductors; if they were, an equalizing current would then flow easily in and between the conducting materials. A static charge is most easily obtained by rubbing a hard nonconducting material against a soft or fluffy nonconductor.

The drawing below illustrates how electrons are displaced from a piece of fur and deposited on a hard rubber rod.

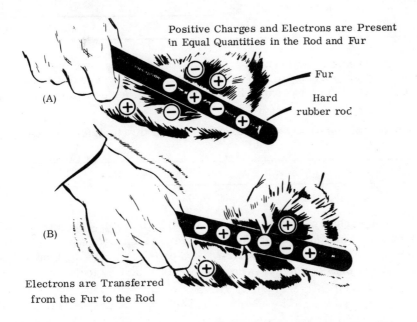

Positive Charges and Electrons are Present
in Equal Quantities in the Rod and Fur

Fur

Hard
rubber rod

(A)

(B)

Electrons are Transferred
from the Fur to the Rod

In the case just illustrated, electrons are transferred from the fur to the rod because they are more easily displaced from the fur. When the hard rubber rod is rubbed with the fur, the rod accumulates electrons. Since both fur and rubber are poor conductors, little equalizing current can flow, and an electrostatic charge is built up. When the charge is great enough, equalizing current will flow regardless of the poor conductivity of the materials. This current may cause a crackling sound, and if it is dark, sparks can be seen.

If a body with a positive charge (too few electrons) comes into contact with a body that has a negative charge (too many electrons), an electric current will flow between the two bodies. Electrons will leave the negatively charged body and enter the positively charged body. The electric current will continue to flow until the charges of the two bodies are equal.

When a body has too many electrons, these electrons do not go into orbit around individual atoms. They are free electrons that give the material an overall negative charge.

When the electric current flows to equalize the charges on the two bodies, static electricity is said to be "discharged." The bodies do not need to touch if the difference between the charges is great enough. An example of this is the lightning that leaps between clouds or between a cloud and the earth during a thunderstorm.

Perhaps you have walked across a carpet and then have touched a bit of metal or even another person, at which time you experienced a slight shock. If so, it is because your body had acquired a negative charge, which could not be dissipated. Later, the shock resulted from the current flow when a touch allowed the charges to equalize. When the static electricity was discharged, the charges were equalized.

What is static electricity? _____

_____

- - - - - - - - - -

Static electricity is the force that exists between unequally charged bodies when current cannot flow between them (or equivalent wording).

30. Try to list one or two other examples of static electricity in everyday life.

_____

_____

- - - - - - - - - -

You can probably think of several. Clothes that cling and crackle when they are removed from an electric dryer are charged with static electricity. And if you have ever rubbed a cat's fur in the dark on a cold night, the sparks you saw were the result of static electricity.

We have seen how the behavior of charged bodies is related to static electricity. We shall examine electric fields and then continue the study of charged bodies in the next two sections.

If you plan to take a break pretty soon, do it now.

Electric Fields

31. The space between and around charged bodies, in which their influence is felt, is called an <u>electric field</u>. (It may also be called an "electro-static field," a "force field," or a "dielectric field.") The field always emanates from material objects and extends between bodies with unlike charges. The fields of force spread out in the space surrounding their points of origin, constantly diminishing as the distance from those points increases.

The field about a charged body is generally represented by lines called <u>electrostatic lines of force</u>. The lines represent the direction and strength of the field. Since a field can exist between a charged body and a neutral body, "positive" can mean "less negative," and "negative" can mean "less positive."

The drawing below represents two pairs of charged bodies. One pair has a positive (+) charge on each body (like charges), while the other pair has a positive charge on one body and a negative (−) charge on the other (unlike charges). The lines of force are indicated in each case.

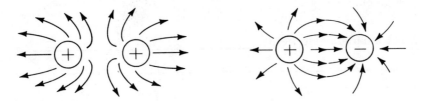

This system of representing lines of force is merely a convention. Although we can measure an electric field, we do not know its exact nature.

Show the charge on each body shown below by drawing the symbol for either positive or negative. (There is more than one correct answer.)

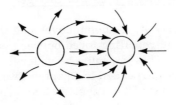

- - - - - - - - - -

One body should be marked positive (+) and the other, negative (−), because unlike charges attract.

32. Show the charge on each body shown below as either positive or negative. (There is more than one correct answer.)

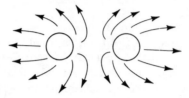

- - - - - - - - - -

You should have labeled both bodies as positive (+) or both bodies as negative (−), because like charges repel.

33. As the electric lines of force travel out into space, does strength of the field increase or decrease? _____

- - - - - - - - - -

It decreases.   (If you weren't sure about this, review frame 31 before you go on.)

34. What is an electric field? _____

_____

- - - - - - - - - -

The space between and around charged bodies in which their influence is felt.   (If you missed this, review frame 31.)

Next we shall study Coulomb's Law of Charges, which describes in more detail the force that exists between charged bodies.

Coulomb's Law of Charges

35. The Frenchman Charles A. Coulomb developed the law that governs the amount of attracting or repelling force between two electrically charged bodies in free space.   This law must be understood before we study electric current, which is introduced in Chapter Two.   Coulomb's formulation, known as Coulomb's Law of Charges, states:

Charged bodies attract or repel each other with a force that is directly proportional to the product of their charges and that is inversely proportional to the square of the distance between them.

This law seems complicated, but it can be broken down into two factors:

    1. The attracting or repelling force depends on the <u>strength of the charges</u> on the two bodies.

    2. It also depends on the <u>distance between the bodies</u>.

Coulomb's Law of Charges takes into account what aspects of any two charged bodies? _____

- - - - - - - - -

The strength of the charges and the distance between the bodies.

36. Coulomb's Law says that the force is directly proportional to the <u>product</u> of the charges on the two bodies. That is, the charge of the first body is multiplied by the charge of the second body. Without worrying about the exact amount of the charges, let's give them number values. The first body has a charge of 2 and the second has a charge of 3. With these numbers assigned, the force is directly proportional to a product of

_____

- - - - - - - - -

2 x 3 = 6.   (If your answer was "5," you were thinking of the <u>sum</u> of the charges, not the product.)

37. If the charges are increased so that their product is 12 instead of 6 (while the distance between the bodies remains the same), the force will be

_____ as much.
    (half/twice)

- - - - - - - - -

twice

38. The force is directly proportional to the product of the charges, but it is <u>inversely</u> proportional to the <u>square</u> of the distance. "Inverse" is the opposite of "direct." In mathematics, an inverse relationship is shown as a reciprocal. For example, 1/4 is the reciprocal of 4, 1/5 is the reciprocal of 5, etc. The inverse of 2 is 1/2. What is the inverse of 4?

_____

- - - - - - - - -

1/4

39. The "square" of the distance merely means the distance multiplied by itself. If the distance between two objects is 3 meters, for example, the square of the distance is 9 meters. If the distance between two objects

is 4 centimeters, what is the square of the distance ?

_____

- - - - - - - - - -

16 centimeters

40. This inverse/square formula is common to all known electromagnetic phenomena. If the distance from a light source, for example, is doubled, the illumination is quartered. Assume that the distance between two charged bodies is 1 centimeter. If the charge on each body remains the same, but the distance is increased to 2 centimeters, the attracting or

repelling force is (1/2, 1/4, 1/8) _____ what it was before.

- - - - - - - - - -

1/4

41. The force between charged bodies can be calculated using Coulomb's Law. However, you are not required to learn the mathematics for doing so in this book. For now, you need only a general understanding of the relationship between charged bodies.

    If the total charge on two charged bodies is increased, and the distance between them remains the same, is the force increased or decreased?

_____

- - - - - - - - - -

It is increased.

42. If the total charge remains the same and the distance between the bodies

is increased, is the force increased or decreased? _____

- - - - - - - - - -

It is decreased.

    In this chapter you have learned the nature of free electrons, on which electric current depends. You have learned why some materials are good conductors and some are not. You have learned the nature of static electricity and how to produce it. You have also learned how charged bodies attract or repel each other. Finally, you have learned how electric fields behave and have been introduced to Coulomb's Law of Charges.

    When you feel you understand all the material in this chapter, turn to the Self-Test.

Self-Test

The following questions will test your understanding of Chapter One. Write your answers on a separate sheet of paper and check them with the answers provided following the test.

1. Name the particle in an atom that has each of the following charges: (a) positive; (b) negative; (c) neutral.

2. What are free electrons?

3. Explain in your own words the flow of electric current in a copper wire when an electric force is applied to the wire.

4. Define a good conductor.

5. Define a good insulator.

6. Give three examples of good conductors.

7. Give three examples of good insulators.

8. State the law of attraction and repulsion of charged bodies.

9. What is static electricity?

10. What is an electric field?

11. According to Coulomb's Law of Charges, what two factors affect the force between two charged bodies?

12. State whether the force between two charged bodies will increase or decrease under each of the following conditions.
    (a) The charge is increased while the distance remains the same.
    (b) The distance is increased while the charge remains the same.

Answers

If your answers to the test questions do not agree with the ones given below, review the frames indicated in parentheses after each answer before you go on to the next chapter.

1. (a) proton, (b) electron, (c) neutron   (2)

2. Electrons that have been forced out of orbit about their original atoms. (6)

3. When an electric force is applied to a copper wire, free electrons are displaced from the copper atoms and move along the wire, producing electric current.   (7)

4. A material that has many free electrons.   (13)

5. A material that has few free electrons.   (14)

6. Any three: silver, copper, aluminum, zinc, brass, iron.   (15)

7. Any three: dry air, glass, mica, rubber, asbestos, bakelite.   (16)

8. Unlike charges attract each other and like charges repel each other.   (22)

9. The force that exists between unequally charged bodies when current cannot flow between them.   (29)

10. The space between and around charged bodies in which their influence is felt.   (34)

11. The strength of the charges and the distance between the bodies.   (35)

12. (a) increase, (b) decrease   (41, 42)

CHAPTER TWO
Understanding Voltage, Current, and Resistance

Throughout most of this book, you will be working with electric circuits. Later in the book, you will learn in detail just what an electric circuit is. For now, you only need to know that it is a complete path for current flow from a battery, or some other source of voltage, through one or more conductors back to the voltage source.

While you will have to understand various circuit factors from time to time, you will almost constantly manipulate values assigned to the three basic circuit factors: voltage, current, and resistance. These three variables are interrelated, so first you must know what they are and how each affects the total electric circuit.

You will also have to know something about magnetism, since alternating-current theory is based on this phenomenon. Magnetism will be discussed in Chapter Six.

When you have finished this chapter, you will be able to:

- relate electromotive force (voltage) to the flow of electric current;

- describe the difference between direct current and alternating current;

- describe the function of resistance in limiting current flow and the factors that affect resistance;

- write and use the symbols representing voltage, current, and resistance;

- describe some general methods of producing voltage; and,

- distinguish between wet-cell and dry-cell batteries and describe their components.

Difference in Potential

1. Just as water pressure causes water to flow in pipes, an electrical "pressure," called a difference in potential, causes current to flow in a conductor. Since a difference in potential causes current flow, you need to understand what this is before you can grasp the concept of current

flow. In the drawing below, part of the water in Tank A will flow into
Tank B when the valve is opened. Draw a line across Tanks A and B to
indicate where you think the water level will be after the water has
stopped flowing.

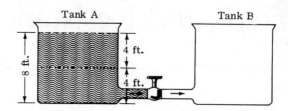

- - - - - - - - -

Your line should be across the approximate middle of both tanks. The
water will flow from Tank A to Tank B until the water level is equal in
the two tanks.

2. Why do you think part of the water flowed into Tank B when the valve

was opened? _____

- - - - - - - - -

The water pressure in Tank A was greater.

3. The "water tank" analogy is useful in understanding one of the basic concepts in electricity: difference in potential. The force that causes free
electrons to move in a conductor as electric current is known as the difference in potential. It is also called electromotive force (emf); but you
are probably familiar with the most common term, voltage. All three of
these terms are interchangeable. In dealing with electricity we naturally
need some units of measurement. The unit of measurement of voltage is
very common and similar to the word "voltage." What do you think that

unit of measurement is? _____

- - - - - - - - -

the volt

4. When a difference of potential exists between two charged bodies that are
connected by a conductor, electrons will flow along the conductor. This
flow is from the negatively charged body to the positively charged body.

Why is this so? _____

_____

- - - - - - - - -

Because electrons are negative and like charges repel, while unlike charges attract.

5. The water stops flowing between two tanks when the pressure on the two tanks is equal. The force with which the water flows, however, is not constant. When do you think that force is greatest? _____

_____

- - - - - - - - - -

You're right if you said: when the pressure differential (difference in pressure) is greatest.

6. The force with which the water flows between the tanks is directly proportional to the pressure differential. Similarly, current flow through an electric circuit is directly proportional to the difference in potential (or voltage) across the circuit. What happens to current flow when the difference in potential is increased? _____

- - - - - - - - - -

Current flow increases.

7. If voltage is increased, current is _____. If voltage is decreased, current is _____.

- - - - - - - - - -

increased; decreased

8. What is the effect on current flow if the voltage is doubled? _____

_____

- - - - - - - - - -

Current flow doubles.

9. "Voltage" is also called _____ or

_____.

- - - - - - - - - -

electromotive force (emf) or difference in potential

10. The abbreviation for "electromotive force" is _____.

- - - - - - - - - -

emf

11. Tell in your own words what is meant by the statement, "Current is directly proportional to voltage." _____

_____

\- \- \- \- \- \- \- \- \- \-

Something like: "When voltage is increased or decreased, current is increased or decreased accordingly."

Electric Current

12. The drift or flow of electrons through a conductor is called electric current or electron flow. Some conventional textbooks make a distinction between current flow and <u>electron</u> flow. Everyone concedes that <u>electron</u> flow is from a negative to a positive terminal, since electrons are negative, and like charges repel. However, some authorities think of <u>current</u> flow as from positive to negative. Since the concepts of "positive" and "negative" are constantly encountered in the theory of electricity, it is important to avoid any confusion about the direction of current flow. In this book, we will make no distinction between current flow and electron flow. The terms current flow and electron flow are interchangeable;

therefore, the direction of current flow is from _____ to

_____.

\- \- \- \- \- \- \- \- \- \-

negative to positive

13. Electric current is generally classified into two general types: direct current (dc) and alternating current (ac). Direct current does not change its direction of flow. Alternating current periodically reverses direction. These two types of current will be discussed thoroughly later in the book. The common battery is a source of direct current. Does the current flow

from a battery change direction? _____

\- \- \- \- \- \- \- \- \- \-

no

14. The current that lights your house changes direction many times per second. What type of current is it? _____

\- \- \- \- \- \- \- \- \- \-

alternating current

15. What is direct current? _____

_____

- - - - - - - - - -

Current that does not change its direction of flow.

16. What is alternating current? _____

_____

- - - - - - - - - -

Current that periodically changes its direction of flow.

17. The greater the voltage, the greater the current flow. Current is measured in <u>amperes</u> (often called "amps" for short). One ampere may be defined as the flow of 6. 28 x 10^{18} electrons per second past a fixed point in
a conductor (but you don't have to know that). Since both quantity and time
are involved, the ampere indicates a <u>rate</u> of current flow. Your house
fuses (or circuit breakers) are rated in amperes; that is, the rate of current flow that will blow the fuse or trip the circuit breaker. A fuse will

blow when too many _____ of current flow in its circuit.

- - - - - - - - - -

amperes

18. The ampere is a measurement of the rate of current flow. The unit that
measures the <u>quantity</u> of electrons is the coulomb, which is defined as
one ampere of current flowing for one second. If we want to measure the
quantity of electrons, rather than the rate of current flow, what unit would

we use? _____

- - - - - - - - - -

coulomb

19. Thus, electromotive force (the "pressure" that causes electrons to move)

is measured in _____. The rate of flow of electric current is

measured in _____. The quantity of electrons is measured in

_____.

- - - - - - - - - -

volts; amperes; coulombs

20. Each of these units has a standard symbol that is used in circuit drawings and equations. The symbol for voltage is the first letter of the phrase, "electromotive force," or E. The symbol for the <u>quantity</u> of electricity, which is measured in coulombs, is Q. The <u>rate</u> of current flow, which is measured in amperes and is very important in the study of electricity, is represented by the symbol I. Write the symbols corresponding to the units of measurement listed below.

    ampere _____

    coulomb _____

    volt _____

- - - - - - - - - -

ampere, I; coulomb, Q; volt, E

21. For each of the following symbols, name the unit represented and tell what the unit measures.

    Q _____

    E _____

    I _____

- - - - - - - - - -

Q, coulomb, measures the quantity of electricity. E, volt, measures the electromotive force. I, ampere, measures the rate of current flow.

Resistance

22. In Chapter One you learned about good conductors and poor conductors (insulators). You learned that free electrons, or electric current, could move easily through a good conductor, such as copper, but that an insulator, such as glass, was an obstacle to current flow. Every material— even copper or silver—offers <u>some</u> resistance, or opposition, to the flow of electric current through it. If the material offers high resistance to current flow, it is termed an insulator. If its resistance to current flow is low, it is called a conductor. The amount of current that flows in a given circuit depends on two factors: voltage and resistance (represented by the symbol R). Thus the amount of current that flows in a circuit can

be changed by changing either _____ or _____.

- - - - - - - - - -

voltage (E); resistance (R)

23. The unit of resistance is the <u>ohm,</u> named for the man who developed Ohm's Law, which you will study in Chapter Three. Resistance, as the word implies, is the ability of a material to impede the flow of electrons. (If you later study advanced electricity, you will encounter another concept of electricity, conductance, which is the opposite of resistance, but you do not need this concept to understand the material in this book.)

    Current is measured in amperes, and voltage is measured in volts.

Resistance is measured in _____.

- - - - - - - - - -

ohms

24. The abbreviation for "volts" is "v." "Amperes" is abbreviated "a." But the small letter "o" looks too much like "a," and "ohm" would be awkward and tiresome to indicate values of resistance. Thus the Greek letter omega (Ω) is used as an abbreviation for "ohms." "Ten amperes" is abbreviated "10 a." Write the abbreviation for "10 ohms" below.

_____

- - - - - - - - - -

   10 Ω

25. Give the abbreviations for the following units of measurement.

      amperes _____

      volts _____

      ohms _____

- - - - - - - - - -

amperes, a.; volts, v.; ohms, Ω

26. The wires that carry current in an electric circuit are usually made of copper, because it is both a good conductor and relatively inexpensive. But the <u>size</u> of the wires is a factor, too. Just as water flows more easily (at a given pressure) in a large pipe than in a small one, electric current flows more easily (at a given voltage) in a large (greater diameter) wire than in a small one. In an electric circuit, at a given voltage, the

larger the diameter of the wires, the (higher/lower) _____

will be the current flow.

- - - - - - - - - -

higher

27. If voltage is held constant, current flow depends on the resistance of the wires (and other devices) in the circuit. A larger diameter wire offers less resistance to current flow than a smaller diameter wire, but one other factor affects the total resistance to current flow: the <u>length</u> of the wires. Since the material of the wires offers resistance to current flow, increasing the amount of material (for a given diameter) increases total resistance. Increasing the length of the wires in an electric circuit has

what effect on resistance? _____

_____ This in turn has what effect on the

amount of current flowing in the circuit? _____

_____

- - - - - - - - - -

It increases resistance; it decreases the amount of current flow.

28. It would not be practical to change the rate of current flow by changing the size or length of the wires. Electrical circuits require varying amounts of current flow for different uses. For that reason, parts are manufactured that present precise amounts of opposition or resistance to current flow. These devices are called, not very surprisingly, <u>resistors</u>. The amount of resistance is measured in <u>ohms</u>. (Although you don't need to know it at this point, one ohm is the resistance in a circuit that permits a steady current of one ampere—one coulomb per second—to flow when a steady emf of one volt is applied to the circuit.) If you want to increase the current flow in a circuit, one way is to take out a resistor and re-

place it with one rated at (more/fewer) _____ ohms.

- - - - - - - - - -

fewer

29. Without changing the wiring or values of resistors, how could you increase

current flow? _____

- - - - - - - - - -

By increasing the voltage applied to the circuit.

30. How does increasing the diameter of the wires affect resistance?

_____

- - - - - - - - - -

Resistance is decreased.

31. Increasing the length of the wires in a circuit has what effect on resistance? _____

- - - - - - - - - -

Resistance is increased.

32. Increasing the voltage applied to a circuit has what effect on current flow? _____

- - - - - - - - - -

It increases current flow.

33. Without changing the wiring, resistors, or other devices in a circuit, how could you decrease current flow? _____

- - - - - - - - - -

Decrease the voltage.

34. Later you will use the symbols E, I, and R in equations to solve problems involving electric circuits, so it is important that you remember them.

The symbol for voltage is _____; for current, _____; and for resistance, _____.

- - - - - - - - - -

voltage, E; current, I; resistance, R

Primary Methods of Producing a Voltage

35. There are many ways to produce electromotive force, or voltage. On the following page, match the six most common methods in the first column with their appropriate descriptions in the second column.

_____ 1. FRICTION   (a) Voltage produced by heating the junction where two unlike metals are joined.

_____ 2. PRESSURE   (b) Voltage produced in a conductor that is moved in a magnetic field.

_____ 3. HEAT   (c) Voltage produced by squeezing certain crystals (piezoelectricity).

_____ 4. LIGHT   (d) Voltage produced by the use of certain photosensitive substances.

_____ 5. CHEMISTRY   (e) Voltage produced in a battery cell.

_____ 6. MAGNETISM   (f) Voltage produced by rubbing two materials together.

- - - - - - - - -

1. (f); 2. (c); 3. (a); 4. (d); 5. (e); 6. (b)

36. In understanding the fundamentals of electricity, you will be most concerned with chemistry and magnetism as means to produce voltage. Friction has little practical application, although we discussed it earlier in studying static electricity. Heat, light, and pressure do have useful applications, but we do not need to consider them in the early stages of study. Chemistry and magnetism, on the other hand, are the principal sources of voltage. Anyone who drives a car is familiar with one of these voltage sources, because if it is dead or missing, the car can't be started. This voltage source, which uses chemistry as its basis, is the

_____.

- - - - - - - - -

battery

37. But the battery alone cannot keep the automobile running. A generator, or alternator, supplies the voltage necessary for running the engine and keeping the battery charged. This device operates by moving a conductor in a magnetic field. Thus, an understanding of both chemical action and magnetism is necessary to understand practical electricity. A generator produces alternating current and uses the principles of magnetism. A battery employs which of the six methods of producing voltage?

_____

- - - - - - - - -

chemistry

Batteries

38. Batteries are widely used as sources of direct-current electrical energy
in automobiles, boats, aircraft, portable electric and electronic equipment, and lighting equipment. A battery consists of a number of <u>cells</u>
assembled in a common container and connected together to function as a
source of electrical power. The ordinary flashlight battery is not actually
a battery but a cell, according to this definition, because two or more
"batteries" (which are actually cells) operate together to provide the pow-

er. Which parts of a flashlight make up the true battery? _____

_____

- - - - - - - - - -

The cells and the barrel of the flashlight, which contains the cells (or
some similar wording). The bulb, lens, and switch are not part of
the battery.

39. The <u>cell</u> is the device that transforms chemical energy into electrical
energy. It consists of a <u>positive electrode</u> (carbon), a <u>negative electrode</u>
(zinc), a chemical solution, and a glass container. The chemical solution
is the <u>electrolyte</u>.
    Aside from the container in which they are assembled, the basic parts

of a simple cell are: two _____, and a chemical solution

called the _____.

- - - - - - - - - -

electrodes; electrolyte

40. The basic device that transforms chemical energy into electrical energy

is the _____.

- - - - - - - - - -

cell

41. A _____ consists of two or more cells assembled
in a common container to produce electricity.

- - - - - - - - - -

battery   (Note: You <u>could</u> have a battery with only one cell, but this
is not usually the case.)

42. What is the relationship between a cell and a battery? _____

_____

- - - - - - - - - -

The battery includes the cells and the container.

43. The cells of a battery are like "mini-batteries." The battery has internal
    electrical connections between the cells, and then the battery provides
    power to some external circuit. Each cell provides part of the power to
    the entire circuit. The electric current that flows in a circuit "powered"
    by a cell consists of free electrons that leave the (negative/positive)

    _____ electrode and return to the cell through the

    (negative/positive) _____ electrode. (Hint: Like
    charges repel.)

    - - - - - - - - - -

    negative; positive   (Note: Electrons are negative, so they are repelled
    by the negative electrode and attracted by the positive electrode.)

44. The simple cell consists of two electrodes suspended in a chemical solu-
    tion called an electrolyte. (The electrodes are usually strips, rods, or
    sheets of two different materials. The most common materials are car-
    bon and zinc, but you don't need to know that for this book.) The battery
    works because of the interaction between the chemical solution and the two
    dissimilar materials. Thus, you might say that chemical energy results

    from the interaction of the two different materials, called _____,

    and the chemical solution, called the _____.

    - - - - - - - - - -

    electrodes; electrolyte

45. Batteries are classified as either wet-cell or dry-cell batteries. The dif-
    ference lies in the form of the electrolyte. The wet cell has a liquid elec-
    trolyte, but the dry cell is not completely dry; its electrolyte is actually
    a damp paste. (If it were completely dry, there would be almost no chem-
    ical action.) A common example of the wet-cell battery is the storage
    battery in an automobile. An example of the dry-cell battery is the com-
    mon flashlight. The major difference between a wet-cell battery and a
    dry-cell battery is in the form of the _____.

    - - - - - - - - - -

    electrolyte

46. What general type of battery has cells whose electrodes are suspended in a liquid electrolyte? _____ Which type has an electrolyte that is a wet paste? _____

- - - - - - - - - -

The wet-cell battery; the dry-cell battery.

47. For most applications, a dry cell is more convenient as the basic unit of a battery. Shown here is a cutaway view of a typical dry cell.

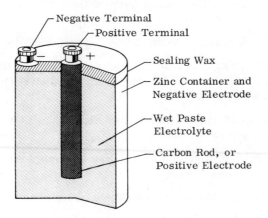

Which part of the cell is the negative electrode? _____

Which part is the positive electrode? _____

Which electrode is buried in the electrolytic paste? _____

- - - - - - - - - -

The negative electrode is the zinc container; the positive electrode is the carbon rod, which is buried in the electrolytic paste.

You are now able to relate electromotive force (difference in potential) to the flow of electric current and to describe how resistance limits current flow. You have become familiar with the symbols representing voltage, current, and resistance. You have learned some general methods of producing voltage as well as the types and composition of one such method: batteries. Now complete the Self-Test.

Self-Test

The following questions will test your understanding of Chapter Two. Write your answers on a separate sheet of paper and check them with the answers provided following the test.

1. One battery terminal is negative and the other is positive. What force of "pressure" causes current to flow in the circuit connecting the terminals?

2. Why is current flow in a battery-powered circuit from the negative terminal to the positive terminal?

3. What is the effect on the rate of current flow if the voltage is reduced by 50 percent?

4. What is the difference between direct current and alternating current?

5. Give the unit of measurement and symbol for each of the following:
   (a) difference in potential; (b) rate of current flow; (c) quantity of electricity; (d) resistance.

6. Give the abbreviation for each of the following units of measurement:
   (a) ampere; (b) volt; (c) ohm.

7. The rate of current flow in a circuit can be changed by changing what other two variables?

8. What three factors affect the resistance in a circuit?

9. Give two ways of increasing the resistance of the circuit wiring.

10. What are the two most widely used methods of producing a voltage?

11. What method of producing a voltage is used in batteries?

12. What is the relationship between a cell and a battery?

13. Current flows away from what electrode of a cell?

14. What are the two basic components that cause a cell to produce electricity?

15. What is the major difference between a wet cell and a dry cell?

16. What is the form of the electrolyte of a dry cell?

Answers

If your answers to the test questions to not agree with the ones given below, review the frames indicated in parentheses after each answer before you go on to the next chapter.

1. Difference in potential.   (1-4)

2. Like charges repel, so the negative terminal repels electrons, which are negative. In addition, the positive terminal attracts the electrons.   (4)

3. The rate of current flow is reduced by 50 percent.   (6-8)

4. Direct current flow does not change direction, while alternating current periodically changes direction.   (13–16)

5. (a) volt, E; (b) ampere, I; (c) coulomb, Q; (d) ohm, R   (17–23)

6. (a) a.; (b) v.; (c) Ω (24-25)

7. Voltage and resistance   (22)

8. The values of the resistors, the length of the conductors, and the diameters of the conductors.   (26-28)

9. Make the wires longer or decrease their diameter.   (30-31)

10. chemistry; magnetism   (36)

11. chemistry   (37)

12. The battery includes the cells and the container.   (42)

13. negative   (43)

14. The electrodes and the electrolyte.   (44)

15. The form of the electrolyte.   (45)

16. A moist paste.   (46)

CHAPTER THREE
The Simple Electric Circuit

In Chapters One and Two you were introduced to some of the basic concepts of electricity. Now you will use those fundamental ideas to begin to understand the electric circuit.

You will need to know some algebra to work through this chapter and many of those that follow. If you need to learn how to solve simple equations (or if you would like a review), turn to Appendix I. The material presented there will teach you all you need to know about the math required in this book.

When you have finished this chapter you will be able to:

- draw schematic diagrams of simple electric circuits, using standard schematic symbols;

- solve simple equations, derived from Ohm's Law, to find voltage, current, and resistance;

- relate power consumption to voltage, current, and resistance, and solve power equations;

- solve problems about the power capacity of electrical devices; and

- relate power to energy and solve energy equations.

The Electric Circuit

Refer to Figure 3-1 on the following page for frames 1 through 7.

1. An electric circuit includes an <u>energy source</u>, some kind of <u>load</u> to dissipate the energy, and a <u>conductor</u> to provide a pathway for current flow. The energy source could be a battery, as in Figure 3-1, or some other means of producing a voltage. The load that dissipates the energy could be a lamp, as in Figure 3-1, a resistor, or some other device that does useful work, such as an electric toaster, a power drill, or a soldering iron. (Of course, a circuit might include a great many separate devices, or loads.) The conductor, which is usually wire, connects all the loads in the circuit to the voltage source to provide a complete pathway for <u>current flow</u>. No electrical device dissipates energy unless current flows through it. Since wires are not perfect conductors, they heat up (dissipate energy), so they are actually part of the load. For simplicity, how-

ever, we usually think of the connecting wiring as having no resistance, since it would be tedious to assign a very low resistance value to the wires every time we wanted to solve a problem.

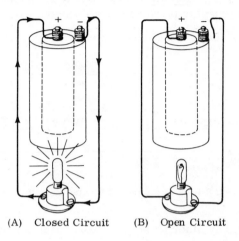

(A) Closed Circuit   (B) Open Circuit

Figure 3-1. (A) Simple electric circuit (closed);
(B) Simple electric circuit (open).

You can check the circuit in Figure 3-1, View A, to see whether there is a complete pathway for current flow. Start at any point, and go around the circuit only once until you return to your starting point. There is no problem until you try to get from one terminal of the battery to the other. There is current flow inside the battery, but here we have a special case, because we said (in Chapter One) that current flows from negative to positive. This is true outside the battery, in what we call the external circuit, because electrons are negative. Since the negative terminal of the battery repels electrons, the current flow is away from the negative terminal and toward the positive terminal (which attracts electrons). The current flow inside the battery (the internal circuit) results from chemical action, not from the laws of attraction and repulsion, and is from the positive terminal to the negative terminal. This completes the circuit. Once you understand this special case of current flow inside the battery, you need not consider it again. Any consideration of current flow in this book is concerned with current flow in the external circuit.

Figure 3-1 shows a very simple electric circuit that includes only a battery, a light bulb, and the connecting wires. Look at the two drawings and note all the differences you see between drawing (A) and drawing (B).

_____

_____

- - - - - - - - - -

(1) The light bulb is lit in (A) but not in (B); (2) arrows are shown along the wire in (A) but not in (B); (3) the wire is connected to the negative (−) terminal in (A) and disconnected in (B).

2. The arrows following the wire, and the "rays" indicating that the bulb is lit, are the artist's way of showing that there is a completed electric circuit in (A) and not in (B).   If there is a complete pathway for the flow of electric current, a <u>closed circuit</u> exists.   If the pathway is interrupted by a break in the conductor (such as a disconnected wire), the result is an <u>open circuit</u>.   Which is the closed circuit, (A) or (B)? _____

- - - - - - - - - -

(A)

3. Which is the open circuit? _____   Why? _____

_____.

- - - - - - - - - -

(B);   Because a wire is disconnected from the negative terminal.

4. Can you think of any other ways (besides cutting the wire) to cause an open circuit in Figure 3-1? _____

_____

- - - - - - - - - -

(1) Disconnect the wire from the positive (+) terminal;   (2) and (3), disconnect a wire from either side of the bulb socket;   (4) unscrew the bulb.

5. The arrows indicate that current flow outside the battery is from negative to positive, but you should know that anyway if you remember the law of attraction and repulsion of charged bodies (Chapter One).   Why is the current flow <u>away from</u> the negative terminal? _____

_____

- - - - - - - - - -

Because electrons, which are negative, are repelled by the negative terminal and attracted by the positive terminal.

6. Each circuit shown in Figure 3-1 can be represented by a <u>schematic diagram</u>.   A schematic diagram (usually shortened to "schematic") is a simplified drawing that represents the <u>electrical</u>, not the physical, situation in a circuit.   Circuit elements are indicated by very simple drawings, called schematic symbols, that are standardized throughout the world, with minor variations.   The following illustrations are the equivalent schematics for the closed-circuit and open-circuit configurations shown in Figure 3-1.

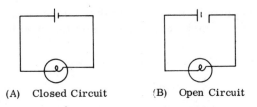

(A) Closed Circuit   (B) Open Circuit

Now, since there are no arrows to show the direction of current flow, there is only one difference between circuit (A) and circuit (B). What is it? _____

- - - - - - - - - -

There is an open space at the top of circuit (B).

7. The drawings in Frame 6 are schematic diagrams representing the elec-trical situation in a circuit, using standardized symbols. (Conductors are simply lines.) The two schematics in Frame 6 show the symbols in the same positions as the circuit elements they represent in Figure 3-1, but this need not be so. A schematic turned on its side or upside down, or with varying line lengths for the conductors, would still be electrically the same. Compare the schematics with the drawings in Figure 3-1, then draw the symbols for a light bulb (lamp) and a battery in the space below.

light bulb   battery

- - - - - - - - - -

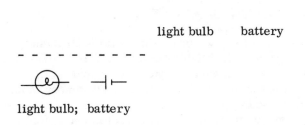

light bulb; battery

8. By convention, the shorter line in the symbol for a battery represents the negative terminal. It is important to remember this, because it is some-times necessary to note the direction of current flow, which is from neg-ative to positive, when you examine a schematic. The battery symbol shown in frame 7 has a single cell, so only one short and one long line are used. The number of lines used to represent a battery vary (and they are not necessarily equivalent to the number of cells), but they are always in pairs, with long and short lines alternating. In the circuit shown in frame 6, the current would flow in a clockwise direction; that is, in the direction that a clock's hands move. If the long and short lines of the battery sym-bol in frame 6 were reversed, the current would flow counterclockwise; that is, in the opposite direction of a clock's hands.

We opened the circuit in Figure 3-1 and in its corresponding schematic (frame 6) by disconnecting a wire from a terminal. Naturally, a switch

is used for this purpose in most circuits. Here is the schematic symbol
for a switch; it may be placed anywhere in the circuit that is convenient.

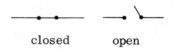

closed   open

Draw a schematic for an open circuit that includes a battery, a lamp,
and a switch. Connect the battery so the current would flow counterclockwise, and label the battery terminals (+) and (−). Use a separate sheet of
paper.

– – – – – – – – – –

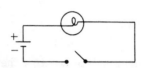

Here is one possibility. The negative side
of the battery is the shorter line. Your
battery can have any number of pairs of
lines. The other schematic symbols may
be shown in any order.

Now that you are familiar with an electric circuit and its equivalent
schematic diagram (other symbols will be introduced as you need them),
you are ready to examine the relationships of current, voltage, and resistance. The relationships are expressed in Ohm's Law, which is the foun-
dation on which electrical theory is based.

Ohm's Law

9. A three-cell flashlight casts a brighter beam than a two-cell flashlight
   using the same size batteries. Flashlights come in different sizes, from
   the little "penlight" on up. Someone had to decide how much voltage a
   cell should produce and what kind of bulb should be used, among other
   things. Flashlight batteries also power toys, clocks, and other devices,
   so circuits have to be designed to produce the correct amount of current.
   How does one decide what the relationships of current, voltage, and resistance should be in a circuit? If the voltage source is constant, how
   much voltage is required to power a device such as a radio or a washing
   machine? This kind of question is basic in electrical design. To answer
   such questions, we start with Ohm's Law.
       Georg Simon Ohm, a 19th century philosopher, formulated the relation-
   ships among voltage, current, and resistance. Ohm's Law states that:

> The intensity of the current in amperes in any electric circuit
> is equal to the difference in potential in volts across the circuit
> divided by the resistance in ohms of the circuit.

You don't have to memorize the law in words, but you <u>do</u> have to memorize the equation that represents it:

$$I = \frac{E}{R}$$

Remember that I represents current expressed in amperes, E represents the difference in potential in volts, and R represents resistance in ohms. What is the effect on current if resistance is increased?

_____

- - - - - - - - - -

Current is decreased.

10. What is another way to decrease current? _____

_____

- - - - - - - - - -

Decrease the difference in potential (voltage).

11. Most of the resistance (sometimes called the load) in a circuit is in the form of components that do specific work, such as a bulb, and certain components, called resistors, whose purpose is to limit current flow. As you have learned, the conductors (wires) themselves have resistance that varies with the size and length of the wire, but it is not practical to limit current flow in this manner. The wire resistance becomes important in advanced electricity, but in this book, you need not consider it. Problems are simplified by ignoring the resistance of the wires. There are no resistors in the simple flashlight, so what component presents most of the

resistance? _____

- - - - - - - - - -

the bulb

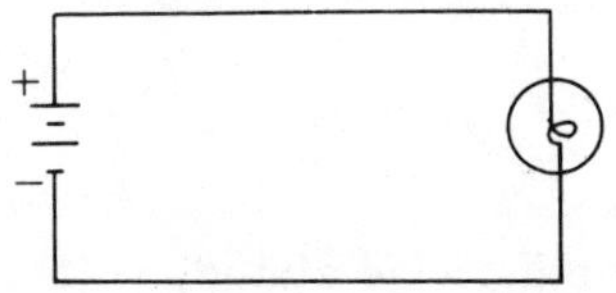

Figure 3-2. Simple electric
circuit with a lamp as the load.

Refer to Figure 3-2 for frames 12 through 25.

12. Assume throughout this book that the wiring has <u>no resistance.</u> To use Ohm's Law to solve for current, voltage, or resistance, you have to know two of the values and solve for the third. To find the current flowing in the

circuit shown in Figure 3-2, you need to know the values for _____

and _____.

- - - - - - - - - -

voltage and resistance

13. Write the equation for finding the current (I) when the voltage (E) and resistance (R) are known. _____

- - - - - - - - - -

$$I = \frac{E}{R}$$

14. The lamp in Figure 3-2 is actually a resistor. The resistance of the lamp is 3 Ω and the battery produces 6 v. How much current is flowing in the circuit? _____

- - - - - - - - - -

2 amperes (2 a.)

Here is the solution: $I = \dfrac{E}{R}$

$$= \frac{6 \text{ v.}}{3 \text{ Ω}}$$

$$= 2 \text{ a.}$$

15. The voltage source (battery) produces 10 v. and the resistance of the lamp is 4 Ω. How much current is flowing in the circuit? _____

- - - - - - - - - -

2.5 a.   (10 ÷ 4 = 2.5)

16. E = 12 v.; R = 4Ω.

I = _____

- - - - - - - - - -

3 a.   (Note: Unless otherwise stated, voltage is assumed to be in volts, resistance in ohms, and current in amperes.)

17. E = 1.5 v.; R = 2Ω.

I = _____

- - - - - - - - - -

0.75 a.

18. By now you know that $I = \dfrac{E}{R}$. If you want to solve for E in terms of I and R, you have to manipulate the equation to isolate E. Do so on a separate sheet of paper.

- - - - - - - - - -

E = IR   Solution: $I = \dfrac{E}{R}$

Cross multiply.

E = IR

(Note: If you had trouble with this, you need to review Appendix A.)

19. Refer again to Figure 3-2. The current in the circuit is 3 a. and the resistance of the lamp is 2Ω. How much voltage is produced by the battery?

_____

- - - - - - - - -

6 v.   (E = IR = 3 a. x 2 Ω = 6 v.)

20. The resistance of the lamp is 1 Ω and the current is 3.73 a. What is the voltage? _____

- - - - - - - - - -

3.73 v.

21. I = 4 a.; R = 1.5Ω.

E = _____.

- - - - - - - - -

E = IR = 4 a. x 1.5 Ω = 6 v.

22. R = 2Ω; I = 0.5 a.

E = _____.

- - - - - - - - -

1 v.

23. E = IR. To solve for R, you must manipulate the equation to isolate R. Do so on a separate sheet of paper.

- - - - - - - - -

$R = \dfrac{E}{I}$ Solution: $E = IR$

Divide each side by I to isolate R.

$$\dfrac{E}{I} = \dfrac{IR}{I}$$

$$\dfrac{E}{I} = R, \text{ or}$$

$$R = \dfrac{E}{I}$$

24. Refer again to Figure 3-2. The battery produces 6 v. and the current in the circuit is 2 a. What is the resistance of the lamp? _____

- - - - - - - - -

$R = \dfrac{E}{I} = \dfrac{6 \text{ v.}}{2 \text{ a.}} = 3 \, \Omega$

25. A 12-volt battery produces a current in the circuit of 0.5 a. What is the resistance? _____

- - - - - - - - - -

$R = \dfrac{E}{I} = \dfrac{12 \text{ v.}}{0.5 \text{ a.}} = 24 \, \Omega$

(Note: The circuit values in these exercises are selected for easy calculation.)

26. You have now used Ohm's Law to solve for all three circuit values: voltage (E), current (I), and resistance (R). It is important to note that resistance (R) is a physical constant. The value of resistance cannot be changed by changing voltage (E) or current (I). Here is an aid to remembering the three basic equations:

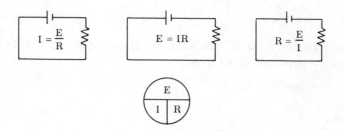

In the preceding circle, cover up the value you wish to solve for, and the rest of the equation is revealed. For example, $I = \dfrac{E}{R}$. Write the equations for E and for R using the memory aid.

- - - - - - - - -

$E = IR; R = E/I$

27. A lamp or any other component has resistance. If the purpose of the schematic is to work out relationships among current, voltage, and resistance, the symbol for a resistor may be used instead of that for the actual component. How much current is flowing in this circuit? _____

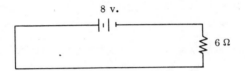

- - - - - - - - -

1.33 a.   (Note: It would be more accurate to say $1\frac{1}{3}$ a., but values are usually given in decimals.)

28. Draw the schematic symbol for a resistor. _____

- - - - - - - - -

—⋏⋏⋏— (Note: Its position does not matter.)

29. The schematic symbol in frame 28 is actually the symbol for a <u>resistance</u> rather than for a resistor only. Any device that consumes power is a resistance. The resistor, which is manufactured to precise specifications and whose function is to limit current, is a resistance, but so is a lamp, an electric iron, or a warming tray. These devices do limit current, and are therefore resistances, but each has a function besides current limiting. When we show the symbol for a resistance in this book, we usually call it a resistor; but remember that the symbol could mean some other kind of resistance.

    Of course, the circuit might include more than one resistor. If so, the resistors are usually labeled R_1, R_2, etc., with their respective values. Resistors in series are merely added to get the total resistance. We shall see why this is true later. In the following drawing, what is the value of R_2? _____

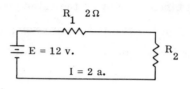

- - - - - - - -

$R_2 = 4\ \Omega$ Solution: $R = \dfrac{E}{I} = 6$

$R_1 = 2\ \Omega$

$R_2 = 6 - 2 = 4\ \Omega$

30. Without changing the resistors in the circuit, how could you increase the current to 4 a. ? _____

- - - - - - - - - -

Substitute a 24-volt battery for the 12-volt battery.
    Solution: $E = IR = 4 \times 6 = 24$

If you need a rest, this is a good place to stop.

Power

31. Perhaps you have noticed a wheel in your home power-meter that turns when electricity is being used. If the main switch is off, the wheel is at rest, because no power is being consumed. As more lights and appliances draw current, the wheel turns faster. Power, whether electrical or me-chanical, pertains to the <u>rate</u> at which work is done, so the power consumption in your home is related to current flow. (You will learn the exact relationship later.) An electric range or dryer consumes more power (and draws more current) in a given length of time than a reading lamp, for example, because more current is required to produce heat than to produce light. (Of course, every device produces some heat.) The am-pere is a measure of the rate at which current flows.   Power is the

_____ at which work is done.

- - - - - - - - - -

rate

32. You might say that an electric range does more work than a reading lamp, but this is true only if the <u>time</u> both appliances are used is the same. If you read a lot and send your clothes to the laundry, your reading lamp could do more work in a year than your dryer. In considering power, we are not concerned with the total amount of work done but the rate at which

it is done. Work is done whenever a force causes motion—even the movement of electrons through a conductor. A dryer uses more current in a
given length of time than a reading lamp, so its power consumption is
greater. State in your own words what it means to say that a dryer consumes more power than a reading lamp. (Hint: Relate your answer to

current flow.) _____

_____

- - - - - - - - - -

A dryer uses more current in a given length of time than a reading lamp.

33. If a compressed spring is held in place between two fixed points, a force
is exerted on those points, but no work is done. Similarly, if a switch is
opened in an electric circuit, a force (difference in potential) exists, but
no work is done because no current flows. When the switch is closed, a
circuit is completed, current flows, and some kind of work is done; a
lamp is lit, perhaps. Only then are we concerned with power. Power is

used only in a(n) (open/closed) _____ electric circuit.

- - - - - - - - - -

closed

34. The basic unit of power, or the rate at which work is done, is the <u>watt</u>.

Just as current is measured in amperes, power is measured in _____.

- - - - - - - - - -

watts   (Note: "Watt" is abbreviated "w.")

35. Power (P) is directly related to the voltage (E) across a circuit and the
current (I) flowing in the circuit. One watt represents the amount of power consumed when the difference in potential of one volt produces a current of one ampere. Power (P) is equal to voltage (E) multiplied by current (I). Write the equation that states this mathematically.

_____

- - - - - - - - - -

$P = EI$ (or $P = E \times I$)

36. A 10-volt battery is the voltage source for a circuit whose total resistance
is 5 Ω. Before you can solve for the power in the circuit, you must first

solve for _____.

- - - - - - - - - -

current (I)   (Note: Later you will be able to solve for power when voltage and resistance are given.)

37. What is the equation for current when voltage and resistance are given?

_____

- - - - - - - - -

$I = \dfrac{E}{R}$

38. What is the current in the circuit described in Frame 36? _____

- - - - - - - - -

2 a.

39. What is the equation for power (P) when voltage (E) and current (I) are given? _____

- - - - - - - - -

P = EI

40. What is the power in the circuit described in Frame 36? _____

- - - - - - - - -

20 w.

41. In Frame 36, E and R were given. But because you knew only the power equation, P = EI, you had to solve for I before you could arrive at P. There is, of course, a short cut.

　　(1) P = EI

　　(2) Substitute the equivalent of I in terms of E and R. You can do
　　　　this, since you know that $I = \dfrac{E}{R}$.

　　(3) $P = E\left(\dfrac{E}{R}\right)$

　　(4) $P = \dfrac{E^2}{R}$

　　　If the voltage in a given circuit is 4 v. and the resistance is 4Ω, what is the power? (Use the equation just developed.) _____

- - - - - - - - -

$$P = \frac{E^2}{R}$$

$$= \frac{4^2}{4}$$

$$= \frac{16}{4}$$

$$= 4 \text{ w.}$$

42. The circuit voltage is 2 v. and the resistance is 12 Ω. What is the power?

_____

- - - - - - - - -

$\frac{1}{3}$ w.   (0.33 w.)

43. E = 3; R = 2.

P = _____

- - - - - - - - -

4.5 w.

44. You can work the mathematics for power when current and resistance are given, without solving first for voltage.

    (1) P = EI

    (2) Substitute the equivalent of E in terms of I and R. Remember that E = IR.

    (3) P = IR(I)

    (4) $P = I^2R$

If the current in a given circuit is 4 a. and the resistance is 10 Ω, what is the power? _____

- - - - - - - - -

$P = I^2R$
  $= 4^2 \times 10$
  $= 16 \times 10$
  $= 160$ w.

45. The circuit current is 0.5 a. and the resistance is 12 Ω. What is the power? _____

- - - - - - - - -

3 w.

46. I = 4 a.; R = 2Ω.

    P = _____

- - - - - - - - - -

    32 w.

47. Write the equation for P when E and I are given. _____

- - - - - - - - - -

    P = EI

48. Write the equation for P when E and R are given. _____

- - - - - - - - - -

    $P = \dfrac{E^2}{R}$

49. Write the equation for P when I and R are given. _____

- - - - - - - - - -

    $P = I^2R$

You probably know that an electric dryer or stove requires a higher voltage than most other appliances in the home. The conversion of electrical energy into heat uses much more current than a radio or electric drill, for example, so some devices require more voltage—and heavier wiring to protect them from damage. Power calculations are quite important in advanced electricity and electronics, but one important application of power theory is discussed in the next section.

If you plan to take a break pretty soon, do it now.

Rating of Electrical Devices by Power

50. When you replace a burned-out light bulb, how do you decide which new bulb to select as a replacement? _____

_____

- - - - - - - - - -

You look on the bulb to read its wattage.

51. Light bulbs, soldering irons, and motors are a few of the electrical de-
    vices that are rated in watts. The wattage rating of a device indicates the
    rate at which the device converts electrical energy into some other form
    of energy, such as heat, light, or motion. An electric lamp converts

    electrical energy into _____.

    - - - - - - - - - -

    light   (Note: Some energy is converted into heat, because the lamp is
    not 100 percent efficient.)

52. In the kitchen, electrical energy is converted into light or into mechanical
    energy (in the case of a blender, for example). If you have an electric

    range, electrical energy is also converted into _____.

    - - - - - - - - - -

    heat

53. The greater the wattage of an electrical device, the greater the rate at
    which electrical energy is converted to another form. A 100-watt bulb

    produces more light than a (50/150) _____ -watt bulb.

    - - - - - - - - - -

    50

54. A soldering iron converts electrical energy into heat. Soldering irons are
    rated in watts. State in your own words the difference between a 300-watt

    and a 500-watt soldering iron. _____

    _____

    - - - - - - - - - -

    The 500-watt iron draws more current and produces more heat.

55. Since power is related to voltage, current, and resistance, electrical de-
    vices produce their rated power only if they are operated at the correct
    circuit values. In a home electrical circuit, only the

    (current/voltage/resistance) _____ is constant.

    - - - - - - - - - -

    voltage

56. Electrical devices are rated for voltage as well as wattage (power). A de-
    vice will draw the proper amount of current only if the correct voltage is
    applied. A light bulb, for example, is designed to produce a certain

amount of light.   To do so, it must be operated at a voltage that will result in the right amount of current, and the proper power consumption, to
produce that much light.   If the applied voltage is too low, the light will
be dimmer than is desired.   If too much voltage is applied, the light will
be brighter, and it will probably burn out, because it draws too much current for its filament to withstand safely.   A light bulb is labeled with its
wattage, such as 100 w. , and also with its proper voltage, which is usually 115 v. (standard house voltage).   If a 115-volt lamp is plugged into a

230-volt circuit, what happens to the current? _____

- - - - - - - - - -

It doubles.

57. Let's see how this wattage and voltage rating works.   A 100-watt lamp is
     rated for 110 v.   When it is turned on, how much current flows in its cir-
     cuit? (Hint: Remember the equation, P = EI.)   In this and other problems,
     you may round off your answer to two decimal places for simplicity.

     _____

- - - - - - - - - -

0. 91 a.   Solution: P = EI

$$\frac{P}{E} = \frac{EI}{E}$$

$$\frac{P}{E} = I$$

$$I = \frac{P}{E} = \frac{100}{110} = 0.909$$

58. If the same 100-watt lamp is plugged into a 220-volt circuit, how much

     current flows? _____

- - - - - - - - - -

1. 82 a.

59. How much power is now consumed by the lamp? (Hint: The lamp is rated
     at 100 w. , but it could consume much more power momentarily before it

     burns out.   E = 220; I = 1.82.) _____

- - - - - - - - - -

400. 4 w.

    Note that the power did not merely double when the circuit voltage was
doubled.   The power quadrupled.   (The answer was not exactly 400 w. because some values had been rounded off earlier.)   That is because not only
the voltage but the current doubled.   The bulb would probably burn out when

the lamp is plugged into the 220-volt circuit. Its life would certainly be shortened. If the normal wattage rating of a device is exceeded (by using an incorrect voltage), it will overheat and will probably suffer damage. Now let's see how electrical devices are rated to prevent damage.

Power Capacity of Electrical Devices

60. The wattage rating of a light bulb indicates its ability to do work. A 100-watt light bulb, for example, is expected to produce a certain amount of light when it is used in its normal circuit. However, the wattage rating of some devices indicates operating limits. A resistor is one such device. It is designed to be used in circuits with widely varying voltages, depending on the desired current. But each resistor has a maximum current limitation for each voltage applied. The product of the voltage drop across a resistor and the current through it (the result when these values are multiplied) must not exceed a certain wattage, since this wattage indicates the maximum safe power consumption. (Remember, $P = EI$.) A 10-watt resistor subject to 10 v. has a maximum current limitation of 1 a. because this combination of voltage and current results in a power consumption of 10 w. If the voltage across the resistor is increased, less current is needed to produce the same power, so the maximum current limitation of the resistor will be less. If the voltage across our 10-watt resistor is 20 v., what is the maximum current that can safely flow through the resistor without damage? _____

- - - - - - - - - -

0.5 a.

61. All the resistors in Figure 3-3 are labeled with their wattage rating.

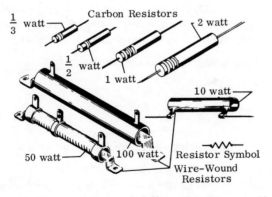

Figure 3-3. Resistors of different wattage ratings.

Small resistors are color-coded for these values; larger resistors are stamped with their resistance in ohms. The two most common types are carbon resistors and wire-wound resistors. Resistors rated at 2 w. and under are usually carbon resistors. What type of resistors are used for higher wattage ratings? _____

- - - - - - - - - -

wire-wound

62. From your examination of the carbon resistors in Figure 3-3, what can you say about the relationship of the size of a resistor to its wattage rating? _____

- - - - - - - - - -

Something like: The larger the resistor, the higher its wattage.

63. Resistors with ratings of 2 watts or less are generally carbon, while wire-wound resistors are usually made in ranges of 2 to 200 watts. (Resistors with higher wattage ratings are of special construction.) Opposite the ratings shown below, indicate the probable construction of a resistor with that rating.

    $\frac{1}{2}$ w. _____   50 w. _____

    1 w. _____   150 w. _____

- - - - - - - - - -

$\frac{1}{2}$ w., carbon; 1 w., carbon; 50 w., wire-wound; 150 w., wire-wound.

64. When current passes through a resistor, electric energy is transformed into heat, which raises the temperature of the resistor. If the temperature becomes too high, the resistor may be damaged. In a wire-wound resistor, the metal wire may melt, opening the circuit and interrupting current flow. This effect is used in devices that protect household (and other) electrical circuits from overloading. From your own experience, what do you think these devices are? (Hint: They will "blow" if too many appliances are plugged into the same outlet.) _____

- - - - - - - - - -

Fuses

65. Figure 3-4, on the following page, includes the symbol for a fuse. It is labeled F_1.

Look at the figure, then draw the schematic symbol for a fuse.

_____

- - - - - - - - - -

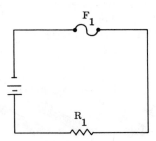

Figure 3-4. Simple circuit that
includes a fuse and a resistor.

66. Fuses are actually metal resistors with very low resistance values. They
are designed to "blow" when the current exceeds a certain value. A circuit that includes a resistor and a fuse is shown in Figure 3-4. Must the
current that flows through the resistor also flow through the fuse? ____

- - - - - - - - - -

yes

67. The fuse (F$_1$) in Figure 3-4 is rated at 0.5 a. and has a resistance of 1Ω .
R$_1$ has a value of 29Ω. The applied voltage is 6 v. What is the current
in the circuit? _____

- - - - - - - - -

0.2 a.   (I = $\dfrac{E}{R}$ = $\dfrac{6}{29 + 1}$ = 0.2)

68. Will the fuse blow? (Yes/No) _____, because _____

_____.

- - - - - - - - - -

No, because the current is less than the rated value for the fuse (which
is 0.5 a.).

69. If R$_1$ in Figure 3-4 has a value of 7Ω and the voltage remains unchanged
(6 v.), what is the circuit current? _____

- - - - - - - - - -

0.75 a.   (6 ÷ 8 = 0.75)

70. Will the fuse blow? Explain. _____

- - - - - - - - - -

Yes, because the current now exceeds the rated value of the fuse.
(Note: The fuse will blow when the current exceeds 0.5 a. The current
is now 0.75, or 50 percent over maximum.)

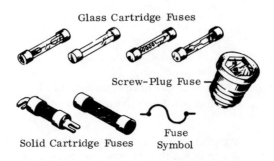

Figure 3-5. Typical fuse types.

Figure 3-5 shows some typical types of fuses. The screw-plug fuse is
commonly used for most house circuits, but solid cartridge fuses are
used for heavier-duty circuits such as those for the kitchen range and the
clothes dryer. Automobile accessory circuits, such as the radio, usually
have glass cartridge fuses. Circuit breakers, which are merely reset,
are now becoming more and more common. The breaker is "thrown" when
the current exceeds the rated value for the circuit breaker.

A subject closely related to power is energy, the subject of the next
section.

Energy

So far in this chapter we have dealt with two basic concepts—current flow
and power. Before we move on to still another related concept—energy—
it might be a good idea to tie together the material covered so far.

All three values included in Ohm's Law (voltage, current, and resistance) are related to one thing: current. Voltage is the "pressure" that
causes current to flow, and resistance is the factor that limits the current
flow when voltage of a given value is applied.

Power, too, is related to current, because no work is done (and therefore, no power is consumed) unless there is current flow. All of the power equations involve current. Either current (I) is directly included in
the equation, or it affects the other values (E and R) in the equation.

Thus, Ohm's Law deals directly with current and the other variables
that affect it (E and R), while power is directly related to the current flow
in a circuit.

71. <u>Energy</u> is defined as the ability to do work. When the archer draws the bow string, for example, the ability to do work is present, but <u>no work</u> <u>is done until the arrow is released</u>. Then the potential for work (energy) is converted to work actually done (power consumption). Energy is expended when work is done, because it takes energy to maintain a force when that force does work. In electricity, energy (W) is equal to the <u>rate</u> at which work is done, or power (P), multiplied by the length of <u>time</u> (t) the rate is measured. Circle the equation below that expresses this mathematically.

$$P = Wt$$

$$W = Pt$$

$$P = \frac{W}{t}$$

- - - - - - - - - -

W = Pt

72. The symbol for energy, W, comes from "watt." Remember that energy is the <u>rate</u> at which work is done, and "rate" implies a time. Miles per <u>hour</u> and feet per <u>second</u> are both rates. In electricity, W will be in watt-hours if time (t) is in hours. If it is expressed in seconds, W will be in

_____.

- - - - - - - - - -

watt-seconds   (Note: Watt-seconds means "watts per second" or "watts times seconds.")

73. An hour is usually too large a measure for calculations in electricity. The second is much more convenient. Let's return for a moment to our basic power equation, P = EI. Assume a very simple circuit in which a

1-volt battery causes current flow of 1 a. P = _____ x _____, or _____ w.

- - - - - - - - - -

P = 1 v. x 1 a., or 1 w.

74. Power is consumed in the circuit only while current is flowing. Since 1 a. of current flowed with 1 v. applied to the circuit, the power consumed was 1 w. But how much <u>energy</u> was responsible for that amount of power consumption in one second? Simply apply the energy equation, W = Pt. The

energy, or rate at which work is done, is 1 _____ - _____.

- - - - - - - - - -

watt-second

75. Another term for "watt-second" is <u>joule</u>, pronounced "jule." In a given circuit, W is calculated to be 300 watt-seconds. Another way to express W in this circuit is 300 _____ .

- - - - - - - - - -

joules

76. You are billed for energy used in your house by kilowatt-hours. ("Kilo" is a prefix that means 1,000; thus, one kilowatt is equal to 1,000 watts.) Just for fun, assume your electric bill shows that you used 10 kilowatt-hours and translate this figure into joules. _____

- - - - - - - - - -

36,000,000 joules, or 36,000 kilo-joules, or 36 mega-joules
(10 x 1,000 x 60 x 60 = 36,000,000)

77. A 10-volt circuit has a total resistance of 4Ω, and current flows for 2 seconds. P = _____

- - - - - - - - - -

$$P = \frac{E^2}{R} = \frac{10^2}{4} = \frac{100}{4} = 25 \text{ w.}$$

78. In the circuit described in Frame 77, W = _____ .

- - - - - - - - - -

W = Pt = 25 x 2 = 50 joules (watt-seconds)

79. E = 20 v.; I = 4 a.; t = 3 seconds.

P = _____   W = _____

- - - - - - - - - -

P = 80 w.; W = 240 joules

80. E = 15 v.; I = 6 a.; t = 2 seconds.

P = _____   W = _____

- - - - - - - - - -

P = 90 w.; W = 180 joules

81. I = 6 a.; R = 2Ω; t = 4 seconds.

W = _____

- - - - - - - - - -

W = 288 joules   (P = I^2R; W = Pt)

82. Figure 3-6 shows the schematic diagram of an unknown device, represented electrically by a single resistor, R_2, enclosed in a dashed-line box. Device X actually consists of several components, but we are interested only in its total resistance. If you can solve for current, power, and energy, you understand the essential concepts presented in this chapter.

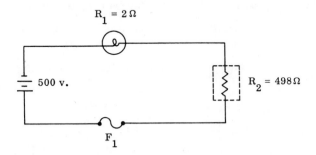

Figure 3-6. Schematic diagram of Device X.

The light, R_1, indicates when the device is running. A timer in Device X opens the switch 5 seconds after power is applied. When the device shuts off, a button must be pressed to restart it.

Assume for simplicity that the fuse and wiring have no resistance.

1. Total current flow is _____.

2. Power consumed during the running time of Device X is _____.

3. The energy developed in the circuit is _____.

- - - - - - - - - -

1. I = 1 a.   (500 v. ÷ 500 Ω)
2. P = 500 w.   (500 v. x 1 a.)
3. W = 2500 watt-seconds, or joules.   (500 w. x 5 seconds)

You have learned to apply Ohm's Law to solve for voltage, current, and resistance. You have built on that basic information to solve for power consumption and energy in a simple electric circuit. You have also become aware that electrical devices can be damaged if the power consumption is beyond their capacity. And you have begun to use the schematic symbols and equations that will be your shorthand as we go further into electrical theory. Now proceed to the Self-Test.

Self-Test

The following questions will test your understanding of Chapter Three. Write your answers on a separate sheet of paper and check them with the answers provided following the test.

1. Draw a schematic diagram of an electric circuit that includes (a) a battery, (b) a resistor, (c) a lamp, (d) a fuse, and (e) a switch in the open position. Show the battery connected so that current would flow counterclockwise in the circuit. Label the battery terminals (+) and (−).

2. Complete these equations, which are developed from Ohm's Law.
   (a) I = ?   (b) E = ?   (c) R = ?

3. Refer to the schematic diagram to solve the following problems.

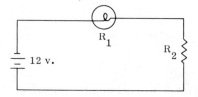

   (a) $R_1 = 3\Omega$; $R_2 = 9\Omega$. I = ?
   (b) $R_1 = 5\Omega$; I = 2a. R_2 = ?
   (c) $R_1 = 7\Omega$; $R_2 = 25\Omega$. To produce a current of 0.75 a., remove the 12-volt battery and replace it with a battery of how many volts?

4. Write the power equation for each set of known values listed: (a) Voltage and current are known; (b) Voltage and resistance are known; (c) Current and resistance are known.

5. Write the equation for energy when power and time are known.

6. Refer to the schematic diagram to solve the following problems.

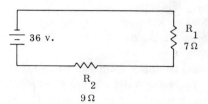

   (a) P = ?
   (b) The current flows for 3 seconds before the circuit is opened. W = ?
   (c) How much energy will be developed in the circuit if a 24-volt battery is substituted for the 36-volt battery and current flows for 3 seconds?

7. How much current is used to illuminate a 100-watt light bulb rated at 115 v. ?

8. A resistor rated at 2 watts is probably wire-wound or carbon?

9. Name two devices used to protect circuits from overloads.

10. What is the difference between power and energy in an electric circuit?

Answers

If your answers do not agree with those below, review the frames in parentheses before going on to the next chapter.

1. Here is one possibility. The negative side of the battery is a shorter line. Your battery can have any number of pairs of lines. The other schematic symbols may be shown in any order.   (1-8, 27-29, 65)

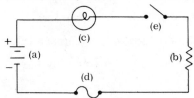

2. (a) $I = \dfrac{E}{R}$ (13)

   (b) $E = IR$ (18)

   (c) $R = \dfrac{E}{I}$ (23)

3. (a) $I = 1a.$ (14)
   (b) $R_2 = 1\Omega$ (25)
   (c) 24   (30)

4. (a) $P = EI$ (35)

   (b) $P = \dfrac{E^2}{R}$ (41)

   (c) $P = I^2R$ (44)

5. $W = Pt$

6. (a) 81 w. You could have used $P = \dfrac{E^2}{R}$; or, solving for current first, $P = EI.$ (41)
   (b) 243 joules (or watt-seconds)   (78)
   (c) 108 joules   (78)

7. 0.87 a. $(P = EI; \; I = \dfrac{P}{E})$ (57)

8. carbon   (61)

9. Circuit breakers and fuses   (66)

10. Power is the rate at which work is done, while energy is the ability to do work.   (31, 71)

CHAPTER FOUR
Series and Parallel Circuits

In Chapter Three you learned Ohm's Law, which is fundamental in electricity. You became familiar with some basic circuit components (battery, lamp, resistor, switch, fuse) and their schematic symbols, and you learned to solve problems involving voltage, current, resistance, power, and energy.

When you finish this chapter you will know how to:

- apply Kirchhoff's Law of Voltages to series circuits;

- apply Ohm's Law to solve for values in parallel circuits;

- apply Kirchhoff's Current Law to solve for current flow in parallel circuits;

- distinguish between series and parallel circuits;

- solve for total resistance in parallel circuits; and

- trace a circuit to establish the polarities of voltages.

Series Circuits

1. In the circuits listed in Chapter Three, you probably noticed that the <u>same current</u> in the circuit flowed through each component, such as a resistor, lamp, or fuse, in that circuit. Those circuits were <u>series</u> circuits. Thus, you could define a series circuit as a circuit in which the current has

   (only one/more than one) _____ path.

   - - - - - - - - - -

   only one

2. You also learned that current cannot flow in an open circuit, since there is no complete path for the current. If a light bulb burns out, its filament breaks, and interrupts the circuit. In strings of old-fashioned Christmas tree lights (and some of the less-expensive lights available today), if one bulb burns out, all lights on the string go out. The lights in such a string

   are part of a _____ circuit.

   - - - - - - - - - -

   series

3. So far you have solved problems involving source voltage only. But when current flows through any device (sometimes called "load") that has resistance, we say that a voltage is "dropped" across the device. (The term "drop" is used because, when voltage is developed in one location, there is a drop in the voltage available at other locations.) In each case, you can calculate the amount of voltage that is dropped across a device if you know its resistance and the current flowing through it, using the formula $E = IR$ that you already know. In the circuit below, solve for the voltage drops across the two resistors.

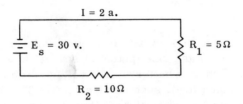

$$E_{R_1} = I \times R_1 = 2 \text{ a.} \times 5\,\Omega = 10 \text{ v.}$$

$$E_{R_2} = \underline{\hspace{2cm}}$$

- - - - - - - - -

$E_{R_2} = 20$ v.   Solution: $E_{R_2} = I \times R_2 = 2$ a. $\times 10\,\Omega = 20$ v.

4. The source voltage (E_S) in the circuit is 30 v.

$$E_{R_1} + E_{R_2} = \underline{\hspace{2cm}}$$

- - - - - - - - -

30 v.

5. From Frame 4 we can deduce (though we won't prove it here) that the total of all voltage drops in a series circuit is equal to the _____.

- - - - - - - - -

source voltage

6. Kirchhoff's Law of Voltages states this truth in different words: <u>The algebraic sum of all the voltages in any complete electric circuit is equal to zero.</u> In a circuit that includes three resistors, there will be a source voltage (E_S) and three voltage drops in the circuit.

To demonstrate:

$E_s = E_1 + E_2 + E_3$

Transposing everything to the left side of the equation:
$E_s - E_1 - E_2 - E_3 = 0$.

This is called the algebraic sum of the terms. In other words, the sum of all <u>positive</u> voltages must be equal to the sum of all _____ voltages.

- - - - - - - - - -

negative

7. For any voltage rise there must be an equal voltage <u>drop</u> somewhere in the circuit. The voltage rise (potential source) is usually regarded as the power supply, such as a battery. The voltage drop is usually regarded as the voltage across a load, such as a resistor. The voltage drop may be distributed across a number of resistive elements, such as a string of lamps or several resistors. However, according to Kirchhoff's Law, the sum of their individual voltage drops must always equal the voltage rise supplied by the power source. In the circuit diagram in Frame 3, the source voltage is 30 v. The voltage drop across R_1 is 10 v., and the voltage drop across R_2 is 20 v. The total voltage drop across the two resistors is 30 v., which is the same as _____.

- - - - - - - - - -

the source voltage

8. If R_1 is taken out of the circuit (and its connecting wires joined), the current will increase because the circuit resistance is lower. You could apply Ohm's Law ($I = \dfrac{E}{R}$) to find the current, then apply another Ohm's Law equation ($E = IR$) to find out the voltage drop across R_2, or E_{R_2}. In this case, however, we don't need to do that, because there is only one resistor in the circuit. Applying Kirchhoff's Law of Voltages instead, we know immediately that E_{R_2} is 30 v. Why? _____

- - - - - - - - - -

Because the source voltage is 30 v.

9. In your own words, state the relationship between all the voltage drops in a series circuit and the source voltage. _____

_____

- - - - - - - - - -

The total of all voltage drops is equal to the source voltage.

10. In a circuit whose source voltage is 12 v. and in which there are only two resistors, there is a voltage drop of 10 v. across one resistor. How much voltage is dropped across the other resistor? _____

- - - - - - - - - -

2 v.   (12 − 10)

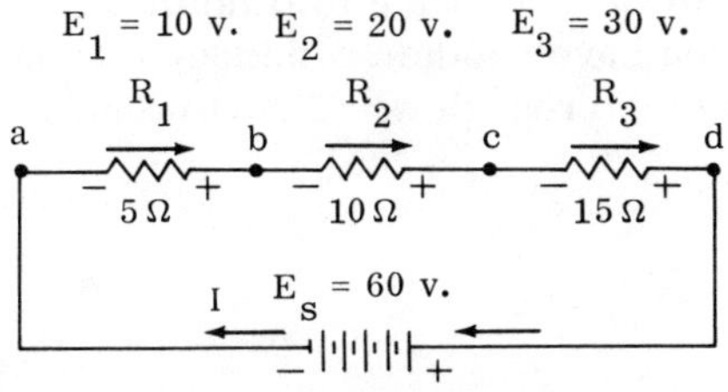

E$_1$ = 10 v. E$_2$ = 20 v. E$_3$ = 30 v.

R$_1$ R$_2$ R$_3$

a   b   c   d

5 Ω   10 Ω   15 Ω

I   E$_s$ = 60 v.

Figure 4-1. Series circuit for demonstrating Kirchhoff's Law of Voltages.

Refer to Figure 4-1 for frames 11 through 20.

11. The "positiveness" or "negativeness" of a value in electricity is called polarity. Whenever there is a difference in potential between two points, such as the two terminals of a battery or the two ends of a resistor, one point is always positive (+) with respect to the other point, which is negative (−). Polarity cannot be assigned to a single point unless that point is compared with some other point. For convenience in understanding the schematic in Figure 4-1, certain points are labeled a, b, c, and d. Look at point b; it is positive with respect to point a but negative with respect to point c. In electricity, we are often interested in the polarity of a difference in potential. Certain devices are designed to be placed in a circuit in a specific way and will not work properly if the polarity is reversed. In troubleshooting electric circuits, the technician often needs to know whether a given difference in potential is positive or negative. E$_1$, the voltage drop across R$_1$, is 10 v. This represents a difference in potential of 10 v. between which two points labeled in the circuit? _____

- - - - - - - - - -

a and b

12. We have not yet assigned a polarity to the difference in potential. We will learn how to do this in a little while, but first let us review voltage drops and current flow. Remember that a voltage drop is a difference in potential. What is the voltage drop between points c and d? _____

- - - - - - - - - -

30 v.

13. What is the voltage drop between points a and c? (Hint: Add E_1 and E_2.)

    _____

    - - - - - - - - -

    30 v.

14. You have been reminded from time to time that current flows from nega-
    tive to positive. You have another reminder in Figure 4-1, since arrows
    show the direction of current flow. The current flow through R_1 is from

    point _____ to point _____.

    - - - - - - - - -

    point a to point b

15. To establish the polarity of a difference in potential, a convention has been
    established that neatly fits the algebraic terms of Kirchhoff's Law of Vol-
    tages. Start at any point in a complete circuit and, following the direction
    of the current flow, label the ends of all loads as positive or negative.
    (A load is any device outside the power supply, such as a resistor or lamp,
    across which there is a difference in potential.) There is also a differ-
    ence in potential across a power supply, and its terminals are labeled (+)
    and (−). Go around the circuit <u>only once</u> and be sure to trace the circuit
    <u>in the direction of current flow</u>. The first point of a load, such as R_1,
    encountered by the current is labeled (−). The <u>other side</u> of that load is
    labeled (+). The end of R_1 in Figure 4-1 encountered second as you trace

    the circuit is labeled (−/+) _____.

    - - - - - - - - -

    +

16. The end of R_2 first encountered as you trace the circuit in the direction

    of current flow is labeled (−/+) _____.

    - - - - - - - - -

    −

17. A question may have occurred to you at this point: Since the right end of
    R_1 and the left end of R_2 in Figure 4-1 are electrically the same, why is
    one point labeled (+) and the other (−)? It is because we are interested in
    the polarity of the <u>voltage across a load</u>, not in the polarity (positive or
    negative) of a single point. By labeling the polarity of the first point en-
    countered when a load is reached, the polarity of the voltage drop across
    the load can be established. Since the first point encountered when you

reach R_1 is negative, E_1 is −10 v. E_2 is (−20/+20) _____ v.

– – – – – – – – –

−20

18. In all work with direct-current electricity, you should label each voltage as either (+) or (−). What is the voltage drop across R_3? _____

– – – – – – – – – –

−30 v.

19. What is the voltage drop across E_s? _____

– – – – – – – – – –

+60 v.   (The first point encountered in tracing the circuit is the positive terminal of the battery.)

20. You can see that the labeling of loads described above is consistent with Kirchhoff's Law of Voltages:

$$E_s + E_1 + E_2 + E_3 = 0$$
$$(+60) + (−10) + (−20) + (−30) =$$
$$+ 60 − 10 − 20 − 30 = 0$$

The algebraic sum of all the voltages in any complete electric circuit is equal to _____.

– – – – – – – – – –

zero

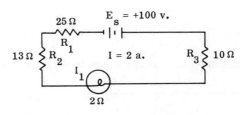

Figure 4-2. A simple DC circuit.

Refer to Figure 4-2 for Frames 21 through 25.

21. A resistor is designated R. What letter designates a lamp? _____

– – – – – – – – – –

I   (Note: Don't be confused by the fact that, in equations, I means current. When a schematic symbol is labeled I, usually with a subscript, it means lamp. Memory aid: I stands for "Incandescent.")

22. Current flow is from E_s toward _____. (Hint: Remember that the short side of the battery symbol indicates the negative terminal.)

- - - - - - - - - -

R_1

23. Label both ends of each load with the proper symbol of polarity (+ or −); then calculate the following voltage drops. (Remember that the current is the same throughout the circuit.)

$E_{R_1} = ?$
$E_{R_2} = ?$

$E_{R_3} = ?$
$E_{I_1} = ?$

- - - - - - - - - -

$E_{R_1} = -50$ v.   (E = IR)

$E_{R_2} = -26$ v.

$E_{R_3} = -20$ v.

$E_{I_1} = -4$ v.

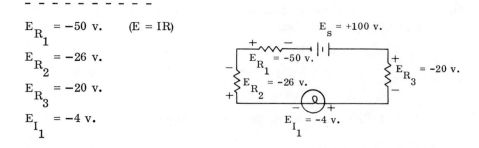

24. What is the sum of the voltage drops across all loads in the circuit outside the battery? _____

- - - - - - - - - -

−100 v.

25. The total voltage drop across all loads in a circuit is the same as the

_____.

- - - - - - - - - -

source voltage, or power supply   (But the polarity is opposite.)

26. From now on, we're going to simplify things a bit. Ignore polarity for this series of frames. Polarities of voltage drops are usually omitted in this book unless they are needed to demonstrate Kirchhoff's Law of Voltages. In many applications in general practice, however, it is necess-

ary to assign polarities. Unlike the practice in algebra, the absence of a sign does <u>not</u> mean (+). Note that we are also simplifying the designations of voltage drops. E_1 is the same as E_{R_1}, E_2 is E_{R_2}, etc. The more complicated subscript is useful, however, when different types of components, such as resistors and lamps, are used as loads. Study Figure 4-3 below and then solve for the voltage drops.

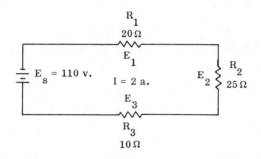

Figure 4-3. Series circuit with three resistors.

Refer to Figure 4-3 for Frames 26 through 39.

$E_1 =$ _____

$E_2 =$ _____

$E_3 =$ _____

- - - - - - - -

$E_1 = 40$ v.   (E = IR)

$E_2 = 50$ v.

$E_3 = 20$ v.

27. What is the total voltage drop across all three resistors? _____

- - - - - - - - - -

110 v.

28. You can establish an equation for the source voltage and all the individual voltage drops in the circuit:

$E_S =$ _____ + _____ + _____

- - - - - - - - -

$E_S = E_1 + E_2 + E_3$

29. To find each voltage drop, you had to apply Ohm's Law, $E = IR$. You knew the total circuit current, which was the same at all points in the series circuit, and you knew each value of resistance. You could have written a variation of the equation in Frame 28:

$$E_s = IR_1 + \underline{\hspace{1cm}} + \underline{\hspace{1cm}}$$

- - - - - - - - - -

$$E_s = IR_1 + IR_2 + IR_3$$

30. R_t is the designation for total circuit resistance. Applying Ohm's Law,

$$E_s = I\underline{\hspace{1cm}}.$$

- - - - - - - - -

$$E_s = IR_t$$

31. Substituting IR_t for E_s, you can write an equation for the source voltage and all the voltage drops in the circuit in terms of I and R:

$$IR_t = \underline{\hspace{1cm}} + \underline{\hspace{1cm}} + \underline{\hspace{1cm}}$$

- - - - - - - - - -

$$IR_t = IR_1 + IR_2 + IR_3$$

32. Thus, you see that E_s can be expressed in terms of circuit current and total circuit resistance: $IR_t = IR_1 + IR_2 + IR_3$. Since there is only one path for current in a series circuit, the total current is the same in all parts of the circuit. Dividing both sides of the voltage equation by the common factor, I, an expression is derived for the total resistance of the circuit:

$$IR_t = IR_1 + IR_2 + IR_3$$

Divide through by I.

$$R_t = \underline{\hspace{1cm}} + \underline{\hspace{1cm}} + \underline{\hspace{1cm}}$$

- - - - - - - - - -

$$R_t = R_1 + R_2 + R_3$$

33. To find the total resistance, R_t, substitute the resistance values in Figure 4-3 and solve for R_t. _____

- - - - - - - - -

$$R_t = 20\,\Omega + 25\,\Omega + 10\,\Omega = 55\,\Omega$$

34. State in your own words the relationship between total circuit resistance and the individual resistances in a series circuit. _____

_____

- - - - - - - - - - -

The total resistance is the sum of the resistances of the individual parts of the circuit.

35. In Figure 4-3, the total resistance is $20 + 25 + 10 = 55\,\Omega$. You can prove this by applying Ohm's Law. E_t designates the total of all voltage drops in the circuit. I_t is 2 a. Using the basic resistance equation $(R = \frac{E}{I})$, solve for R_t. _____

- - - - - - - - - -

$$R_t = \frac{E_t}{I_t}$$
$$= \frac{40 + 50 + 20}{2} = \frac{110}{2}$$
$$= 55\,\Omega$$

36. Now let's solve some power problems using known values from Figure 4-3. The power equation to be used when resistance and current are known is $P = I^2R$. Applying this equation to R_1 in Figure 4-3, we have $P = 2^2 \times 20 = 4 \times 20 = 80$ w., the power absorbed by R_1. How much power is absorbed by R_2? _____

- - - - - - - - - -

100 w.   ($2^2 \times 25$)

37. How much power is absorbed by R_3? _____

- - - - - - - - - -

40 w.

38. What is the total power absorbed by the three resistors in Figure 4-3?

_____

- - - - - - - - - -

220 w.   (You may either add separate power values, or add the resistances and use the equation $P = I^2R$.)

39. Since you know the circuit current and the total of the voltage drops in the circuit, you could have applied another power equation, $P = EI$, or in this case, $P_t = E_t I_t$.

$$P_t = \underline{\hspace{1cm}} \times \underline{\hspace{1cm}} = \underline{\hspace{1cm}} \text{ w.}$$

- - - - - - - - - -

$P_t = 110 \times 2 = 220$ w.

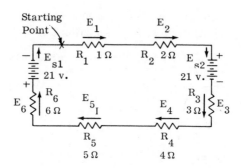

Figure 4-4. Series source
with aiding voltage sources

Refer to Figure 4-4 for Frames 40 through 58.

40. So far the source voltage has been given. Let's look at a case in which two voltage sources are used in the same series circuit. Now we have to work out an effective source voltage to use in solving other problems. If the polarities are such that they aid each other (each causes current to flow in the same direction), the sources are simply added and treated as a single source in solving circuit problems. In Figure 4-4, there are two batteries. Examine the circuit carefully. Is the current flow caused by

E_{s1} in the same direction as the current flow caused by E_{s2}? \underline{\hspace{2cm}}

- - - - - - - - - - -

yes

41. $E_{s1} = $ \underline{\hspace{1.5cm}} $E_{s2} = $ \underline{\hspace{1.5cm}}

- - - - - - - - - - -

Both values are 21 v.

42. What value of source voltage should you use in solving circuit problems?

(Hint: Add E_{s1} and E_{s2}.) \underline{\hspace{2cm}}

- - - - - - - - - - -

42 v. $(E_{s1} + E_{s2} = 21$ v. $+ 21$ v. $= 42$ v. $)$

43. From Figure 4-4, $R_t = $ _____, $I_t = $ _____.

- - - - - - - - - -

$R_t = 21\,\Omega$ (Add the six resistances.)

$I_t = 2$ a. $(I_t = \dfrac{E_s}{R_t} = \dfrac{42}{21} = 2)$

44. $E_4 = $ _____

- - - - - - - - -

8 v. $(E = IR_4 = 2 \times 4 = 8)$

45. $E_6 = $ _____

- - - - - - - - -

12 v.

46. $E_t = $ _____

- - - - - - - - -

42 v. $(2 + 4 + 6 + 8 + 10 + 12 = 42)$

47. The power absorbed by R_3 is _____.   (Hint: $P = I^2R$.)

- - - - - - - - - -

12 w.

48. $P_t = $ _____

- - - - - - - - -

84 w.

49. In Figure 4-4, the two batteries were <u>aiding</u>. Sometimes when a series circuit has more than one voltage source, they might be connected <u>opposing</u> each other to achieve a specific net voltage. If the voltage sources are connected so that each would cause current to flow in a different direction (as they are in Figure 4-5, on the following page), the voltage sources are <u>opposing</u>. Of course, current in a series circuit cannot flow in two directions at once, so the larger source voltage determines the direction of current flow.

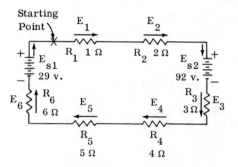

Figure 4-5. Series circuit with
opposing voltage sources.

Refer to Figure 4-5 for Frames 49 through 57.

In Figure 4-5, which is the larger voltage source, E_{s1} or E_{s2}? ____

- - - - - - - - - -

E_{s2}

50. In Figure 4-5, current flow is (clockwise/counterclockwise) _____

_____.

- - - - - - - - - -

clockwise

51. In solving circuit problems involving opposing voltage sources, we have
to work out an <u>effective</u> source voltage. To do this, we must trace the
circuit to find out the polarity of each voltage source. E_{s1} is (+29/−29)

_____ v.

- - - - - - - - - -

−29   (In tracing the circuit, we reach the <u>negative</u> terminal of E_{s1} first.)

52. E_{s2} is (+92/−92) _____ v.

- - - - - - - - - -

+92 v.   (We reach the <u>positive</u> terminal first.)

53. The effective source voltage in Figure 4-5 is the algebraic sum of E_{s1}
and E_{s2}. The effective source voltage is _____ v.

- - - - - - - - - -

+63 v.   (+92 − 29)

54. Once the effective, or net, source voltage is found, circuit problems are solved in the same way as for a single voltage source. We merely regard the effective voltage as the only source voltage. Refer to Figure 4-5 to find the following values.

$$R_t = \underline{\hspace{2cm}}$$

$$I_t = \underline{\hspace{2cm}}$$

$$E_5 = \underline{\hspace{2cm}}$$

- - - - - - - - -

$R_t = 21\,\Omega$ (Add all six resistances.)

$I_t = \dfrac{E_{ff}}{R_t} = \dfrac{63}{21} = 3$ a.

$E_5 = I_t R_5 = 3 \times 5 = 15$ v.

55. The total of all voltage drops across the six resistors is _____ v.

- - - - - - - - - -

63

56. $E_{s1} + E_{s2} + E_1 + E_2 + E_3 + E_4 + E_5 + E_6 = \underline{\hspace{2cm}}$.   (Hint: Remember to assign polarities to all voltages.)

- - - - - - - - - -

zero

57. $P_t = \underline{\hspace{2cm}}$

- - - - - - - - - -

189 w.   (P = EI = 63 x 3 = 189)

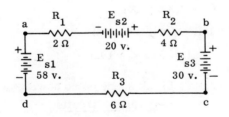

Figure 4-6. Series circuit with three batteries.

Refer to Figure 4-6 for Frames 58 through 61.

58. Even in a more complex problem the principles are the same. In Figure 4-6, the three voltage sources (E_{s1}, E_{s2}, and E_{s3}) are connected so that two are aiding and the third is opposing. Which two voltage sources are aiding? _____

- - - - - - - - - -

E_{s1} and E_{s2}

59. What is the direction of current flow? (Hint: Are the combined voltages E_{s1} and E_{s2} larger or smaller than E_{s3}?) _____

- - - - - - - - - -

counterclockwise   (E_{s1} and E_{s2} combined are larger than E_{s3}, so they establish the direction of current flow.)

60. What is the effective source voltage? _____

- - - - - - - - - -

+48 v.   (Add algebraically the three source voltages: +58, −30, and +20.)

61. Study Figure 4-6 and answer the following questions.

      (a) What is the current in the circuit? _____

      (b) What is the voltage drop between point c and point d? _____

      (c) How much power is absorbed by R_2? _____

      (d) What is the total power consumption in the circuit? _____

- - - - - - - - - -

(a) 4 a. $(I_t = \dfrac{E_s}{R_t} = \dfrac{48}{12} = 4)$

(b) 24 v. $(E = IR = 4 \times 6 = 24)$

(c) 64 w. $(P = I^2R = 4 \times 4 \times 4 = 64)$

(d) 192 w. $(P_t = E_s I_t = 48 \times 4 = 192)$

If you plan to take a break soon, do so now.

Parallel Circuits

62. In a series circuit there is only one path for current flow. As additional loads (such as resistors) are added to the circuit, the total resistance increases and the total current decreases. This is <u>not true</u> in a <u>parallel</u> circuit, as we shall see. In a parallel circuit, each load (or branch) is connected directly across the voltage source. In Figure 4-7, on the next page, the total current flows from the negative terminal of the voltage

source E_S, splits into separate paths at point a, and comes together again at point b.

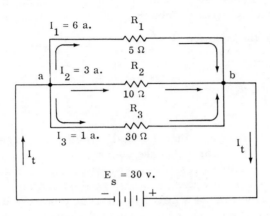

Figure 4-7.   Resistors in parallel.

Refer to Figure 4-7 for Frames 62 through 70.

Starting at the negative terminal of the battery in Figure 4-7, how many separate paths for current flow can you trace? _____ (Remember to go along any path only <u>once</u>.)

- - - - - - - - - -

three

63.   A parallel circuit is one in which there is (only one/more than one) _____ _____ path for current flow.

- - - - - - - - - -

more than one

64.   The total current (I_t) is the sum of all the separate currents in the three branches.   Later we will learn how to solve for the branch currents.   In this case, they are given for simplicity.   What is I_t? _____

- - - - - - - - - -

10 a.   ($I_1 + I_2 + I_3 = 6 + 3 + 1 = 10$)

65.   Examine Figure 4-7.   What is the value of E_s? _____

- - - - - - - - - -

30 v.

66. The total resistance (R_t) in this __parallel__ circuit is __not__ the sum of R_1, R_2, and R_3! Since we already know E_s and I_t, we can apply Ohm's Law to find R_t. Use the equation $R = \dfrac{E}{I}$. $R_t =$ _____ .

- - - - - - - - - -

$$R_t = \frac{E_s}{I_t} = \frac{30}{10} = 3\,\Omega$$

67. You may think it strange that the total circuit resistance is __less__ than that of the smallest resistor. It makes sense, however, if you will consider an analogy with water pressure and water pipes. Let's assume there is some way to keep the water pressure constant. A small pipe offers more resistance to the flow of water than a larger pipe; but if you add another pipe in parallel, even one of very small diameter, the total resistance to water flow is __decreased__. In an electrical circuit, even a larger resistor in another parallel branch provides an __additional path for current flow__, so the __total__ resistance is less. (We proved this in Frame 66.) It is probably more accurate to call this total resistance something like "equivalent resistance," but by convention R_t, or total resistance, is used. If we add one more branch to a parallel circuit, the total resistance (increases/decreases) _____ and the total current (increases/decreases) _____ .

- - - - - - - - - -

Total resistance decreases; total current increases.

68. The voltage across all branches of a parallel circuit is the same, because all branches are connected directly to the voltage source. Consequently, the current through each branch is independent of the others and depends only on the resistance of that branch, as long as the source voltage remains the same. The currents are already indicated in Figure 4-7. However, you can compute them yourself, since the source voltage (the same for all branches) and the resistance of each branch are both known:

$$I_1 = \frac{E_s}{R_1} = 30 \div 5 = 6 \text{ a.}$$

$$I_2 = \frac{E_s}{R_2} = 30 \div 10 = 3 \text{ a.}$$

$$I_3 = \underline{\hspace{1cm}} = \underline{\hspace{1cm}} = \underline{\hspace{1cm}} \text{ a.}$$

- - - - - - - - - -

$$I_3 = \frac{E_s}{R_3} = 30 \div 30 = 1 \text{ a.}$$

69. The total current, I_t, is the sum of all the currents in the parallel branches. This, in somewhat different words, is Kirchhoff's Current Law, which is discussed later in this chapter. Total current is equal to the

sum of the _____ .

- - - - - - - - - -

branch currents (or similar wording)

70. You could work out R_t of a parallel circuit by Ohm's Law, but this would be time-consuming, because you would first have to work out each branch current to arrive at I_t. To arrive at the formula for R_t, you must manipulate equations. As you have seen, I_t is the sum of I_1, I_2, and I_3. $I_1 = \frac{E_s}{R_1}$, etc. Here is the equation for total current I_t (the sum of the branch currents) expressed in terms of E and R:

$$\frac{E_s}{R_t} = \frac{E_s}{R_1} + \frac{E_s}{R_2} + \frac{E_s}{R_3}$$

This introduces R_t, the quantity you are trying to determine, into the equation. Since E_s appears as the numerator in all four factors, it is divided out, and the equation then contains only the desired factor R_t:

$$\frac{1}{R_t} = \frac{1}{R_1} + \frac{1}{R_2} + \frac{1}{R_3}$$
$$\frac{1}{R_t} = \frac{1}{5} + \frac{1}{10} + \frac{1}{30}$$
$$\frac{1}{R_t} = 0.2 + 0.1 + 0.033$$
$$\frac{1}{R_t} = 0.333 \text{ (approximately)}$$
$$R_t = \frac{1}{0.333}$$
$$R_t = \underline{\hspace{1cm}}$$

- - - - - - - - -

3Ω (Note: It is more convenient to work with decimals rather than fractions. The difference is insignificant.)

71. Note that you must work with <u>reciprocals</u> in solving for R_t in parallel circuits. A reciprocal is an inverted fraction; the reciprocal of the fraction $\frac{4}{5}$, for example, is $\frac{5}{4}$. We consider a whole number to be a fraction with 1 as the denominator, so the reciprocal of a whole number is that number

divided into 1. For example, the reciprocal of R_t is $\dfrac{1}{R_t}$. What is the reciprocal of 20 ? _____

- - - - - - - - -

$\dfrac{1}{20}$

72. A parallel circuit has three branches whose resistances are 2Ω, 4Ω, and 10Ω. Solve for R_t. (Give your answer in decimals.)

_____

- - - - - - - - -

1. 176Ω (1.18)   Solution: $\dfrac{1}{R_t} = \dfrac{1}{2} + \dfrac{1}{4} + \dfrac{1}{10}$

$\dfrac{1}{R_t} = 0.5 + 0.25 + 0.10$

$\dfrac{1}{R_t} = 0.85$

$R_t = \dfrac{1}{0.85}$

$R_t = 1.176\Omega$

73. You can see that R_t is always less than the smallest resistance of any branch. A parallel circuit has two branches whose resistances are $1\,\Omega$ and $1,000,000\,\Omega$. R_t is (less/more) _____ than 1Ω.

- - - - - - - - -

less

74. Solving for R_t of a parallel circuit can be sheer drudgery, with many opportunities for arithmetical errors. Fortunately, two shortcuts may be used in certain cases. The first applies only when the parallel resistors (any number of them) all have the same value of resistance. In this case, R_t is found simply by dividing the resistance of one branch by the number of equal branches. Here is the solution for R_t of a parallel circuit that has five branches, each consisting of a 10-ohm resistor:

$R_t = 10 \div 5 = 2\,\Omega$

What is R_t of a parallel circuit that has four branches of 20Ω each?

_____

- - - - - - - - -

5Ω $(20 \div 4)$

75. What is R_t of a parallel circuit consisting of four branches of $16\,\Omega$ each?

_____

- - - - - - - - -

$4\,\Omega$

76. The second shortcut may be used when two and <u>only two</u> branches are connected in parallel. This is called the "product over sum" shortcut, because the product of the two resistances is divided by their sum. Here is
the solution of R_t when two branches have resistances of $3\,\Omega$ and $6\,\Omega$:

$$R_t = \frac{R_1 R_2}{R_1 + R_2} = \frac{3 \times 6}{3 + 6} = \frac{18}{9} = 2\,\Omega$$

Solve for R_t in a parallel circuit where two branches have resistances

of $4\,\Omega$ and $12\,\Omega$. $R_t =$ _____

- - - - - - - - -

$3\,\Omega$ $(4 \times 12) \div (4 + 12)$

77. Two parallel branches have resistances of $5\,\Omega$ and $20\,\Omega$. $R_t =$ _____.

- - - - - - - - -

$4\,\Omega$

Kirchhoff's Current Law Applied to Parallel Circuits

78. Kirchhoff's Current Law states that: <u>At any junction of conductors the
algebraic sum of the currents is zero</u>. This is another way of saying that
as many electrons leave a junction as enter it. Refer once more to Figure 4-7, appearing in Frame 62. Assume that the current flowing toward
junction a (I_t) is positive and the currents flowing away from junction a
$(I_1, I_2,$ and $I_3)$ are negative. (You could assume opposite polarities; it
is only important that the polarity assigned to current flowing <u>toward</u> a
point is opposite to the polarity of any current flowing <u>away from</u> that
point.) Kirchhoff's Current Law is then expressed mathematically:

$$+ I_t - I_1 - I_2 - I_3 = 0$$
$$+ 10 - 6 - 3 - 1 = 0$$

I_t in a parallel circuit with three branches is 12 a. The currents in
the three branches are 7 a., 3 a., and 2 a. Express mathematically this

situation, applying Kirchhoff's Current Law. _____

- - - - - - - - -

$+ 12 - 7 - 3 - 2 = 0$

79. In a three-branch parallel circuit, $I_t = 10$ a., $I_1 = 2$ a., and $I_2 = 3$ a. What is the value of I_3? _____

- - - - - - - - - -

$I_3 = 5$ a.   Solution: $I_t - I_1 - I_2 - I_3 = 0$
$I_3 = I_t - I_1 - I_2$
$I_3 = 10$ a. $- 2$ a. $- 3$ a.
$I_3 = 5$ a.

80. As in the series circuit, the total power consumed in a parallel circuit is equal to the sum of the power consumed in the individual resistors. All the power equations may be applied exactly as they were in the series circuit. For example, the current through R_1 in Figure 4-7 is 6 a., and the voltage drop across R_1 is 30 v. (Remember that the voltage drop across each branch of a parallel circuit is the same as the source voltage.) Here is the solution for power consumed by R_1, using both $P = EI$ and $P = I^2R$:

   (1) $P = EI = 30$ v. x 6 a. $= 180$ w.

   (2) $P = I^2R = 36$ a. x $5\Omega = 180$ w.

   Solve for the power consumed by R_2 using both equations.

   (1) $P = EI$ _____

   (2) $P = I^2R$ _____

- - - - - - - - - -

   (1) $P = 30$ v. x 3 a. $= 90$ w.; (2) $P = 9$ a. x $10\Omega = 90$ w.

81. You found that R_t in Figure 4-7 is 3Ω and I_t is 10 a. Solve for P_t using the equation $P = I^2R$. $P_t =$ _____

- - - - - - - - - -

$P_t = I_t^2R_t = 100$ a. x $3\Omega = 300$ w.

82. Solve for P_t using the equation $P = EI$. $P_t =$ _____

- - - - - - - - - -

$P_t = E_sI_t = 30$ v. x 10 a. $= 300$ w.

83. Problems involving resistance, voltage, current, and power are no more difficult for parallel circuits than for series circuits, although some of the computations take longer. However, it is more difficult to picture the electrical situation in a parallel circuit. If you are given circuit values for a series circuit, you can usually solve any problem associated with that circuit without trouble, simply because there is only one path for current flow. But in solving problems in parallel circuits, you should

always draw a schematic diagram of the circuit, label components, and assign known circuit values. Most people become hopelessly lost without a schematic to help them keep track of known values as well as the unknowns.

Here is a procedure that will help you to visualize what is known and what you need to do to solve for unknowns:

1. Draw a circuit diagram.
2. Write the given values on the diagram.
3. Write down the values to be found.
4. Write the applicable equations.
5. Substitute the given values and solve for the unknown in each equation.

A circuit consists of a 15-ohm and a 35-ohm resistor connected in parallel across a 100-volt battery. Draw the circuit diagram and label the components. Use a separate sheet of paper.

- - - - - - - - - -

Your circuit diagram should look something like this, although your labeling of components might be different. The physical location of components does not matter as long as the electrical connections are correct.

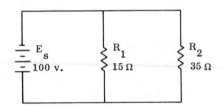

84. Refer to your diagram for the next few frames. It is helpful to write each new value on the diagram as you solve it. R_t = _____. (Hint: You can use the "product over sum" shortcut.)

- - - - - - - - - -

$R_t = 10.5\,\Omega$

85. I_t = _____. (Round off your answer to two decimal places.)

- - - - - - - - - -

9.52 a.

86. The current through R_1 = _____

- - - - - - - - - -

6.67 a.

87. The current through R_2 = _____

- - - - - - - - - -

2. 86 a.

88. The power consumed by $R_1 = $ _____.

- - - - - - - - -

667 w., 666.67 w., or 667.34 w., depending on the power equation you used.

89. The power consumed by $R_2 = $ _____.

- - - - - - - - - -

286 w.   (Some variation, depending on the equation used.)

90. $P_t = $ _____

- - - - - - - - - -

952 w.   (Some variation, depending on the equation used.)

91. Now let's put everything together. Draw your own schematic diagram, using the following information. Three resistors are connected in parallel across a 50-volt battery, labeled E_s for "source voltage." The values of the resistors are:

$$R_1 = 25\,\Omega$$
$$R_2 = 10\,\Omega$$
$$R_3 = 50\,\Omega$$

When you have drawn your schematic and have labeled all circuit components with their values, answer the following questions.

1. What is the total resistance (R_t) of the circuit? _____

2. What is the total current (I_t)? _____

3. What are the three branch currents, I_1, I_2, and I_3? _____

4. What is the total power (P_t) consumed by the circuit? _____

- - - - - - - - - -

Your schematic should look something like this, although the components need not be in the order shown.

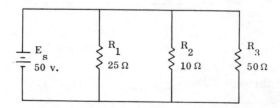

1. $R_t = 6.25\Omega$ $\left(\dfrac{1}{R_t} = \dfrac{1}{R_1} + \dfrac{1}{R_2} + \dfrac{1}{R_3}\right)$

2. $I_t = 8$ a. $\left(I_t = \dfrac{E_s}{R_t}\right)$ You could have solved for the branch currents
first and then added the three.

3. $I_1 = 2$ a.; $I_2 = 5$ a.; $I_3 = 1$ a. $\left(I = \dfrac{E_s}{R}\right.$ for each branch.)

4. $P = 400$ w. $(P_t = E_s I_t.)$ You could have found the power of each branch
   $(P = I^2 R)$ and then added the three branch powers.

In this chapter you have learned that all the voltage drops in a series
circuit must have a total value that is equal to the value of the source voltage although opposite in polarity (Kirchhoff's Law of Voltages). Since
there is only one path for current flow, however, the value of current is
the same in any part of a series circuit.

In a parallel circuit, the total current splits into the various branch
currents, whose sum must equal the total current (Kirchhoff's Current
Law). However, the source voltage is impressed across each branch, so
all branch voltage drops have the same value as the source voltage.

In other words, current is constant and voltage drops vary in a series
circuit, while in a parallel circuit, voltage is constant and current values
vary.

You have learned to trace a series circuit to establish the polarities of
voltage drops in the circuit, in accordance with Kirchhoff's Law of Voltages.

When a series circuit has more than one voltage source, the voltages
either aid or oppose, depending on how they are placed in the circuit, and
you must work out a net voltage to represent E_s. (The same principle applies in parallel circuits. We did not deal with multiple voltage sources
in those circuits because the solution of such problems is too difficult for
this stage of study.)

You have learned that total resistance (R_t) in a parallel circuit requires
a special equation (although two shortcuts may be derived from it for certain cases).

Finally, you have learned to solve for values of current, resistance,
voltage, and power in both series and parallel circuits.

In this chapter, you have encountered circuits that were purely series
or purely parallel. In "real life," of course, circuits usually have portions that are series and other portions that are parallel. We shall examine these circuits in the next chapter.

Now, go on to the Self-Test.

Self-Test

These questions will test your understanding of Chapter Four. Write your
answers on a separate sheet of paper and check them with the answers provided following the Self-Test.

1. What is the difference between a series and a parallel circuit?

2. The sum of all voltage drops in a series circuit is equal to _____.

3. In a series circuit with three resistors, E_s = _____ + _____ + _____.

4. Assuming a series circuit with a battery and two resistors, express Kirch-hoff's Law of Voltages mathematically in terms of the circuit voltages.

5. For any total voltage rise in a circuit, there must be an equal total ____.

Refer to the diagram below for questions 6 through 11.

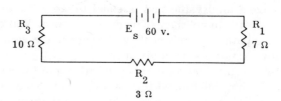

6. Is the direction of current flow clockwise or counterclockwise?

7. Is the voltage drop across R_1 positive or negative?

8. I_t = ?

9. E dropped across R_1 = ?

10. What is the power absorbed by R_2?

11. P_t = ?

Refer to the diagram below for questions 12 through 18.

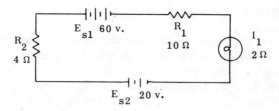

12. Is the direction of current flow clockwise or counterclockwise?

13. What is the effective source voltage?

14. I_t = ?

15. What is the voltage drop across the lamp?

16. What power is consumed by the lamp?

17. What is the current through R_2?

18. P_t = ?

19. As more parallel resistances are added to a circuit, does total resistance increase or decrease?

Refer to the diagram below for questions 20 through 24.

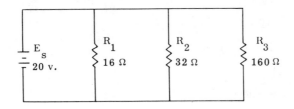

20. R_t = ?

, 21. I_t = ?

22. How much current flows through R_2?

23. What is the power consumed by R_3?

24. P_t = ?

25. Four 5-ohm resistors are connected in parallel across a battery. What is R_t?

26. Two resistors, whose values are 15 Ω and 30 Ω, are connected in parallel across a battery. What is R_t?

Answers

If your answers to the test questions do not agree with the ones given below, review the frames indicated in parentheses after each answer before you go on to the next chapter. (In many cases there is no one specific reference. In these cases, the appropriate equations are given.)

1. A series circuit has only one path for current flow, while a parallel circuit has more than one path. You might also have mentioned that in a parallel circuit, voltage is constant and current values vary. (1, 80, 109)

2. source voltage (5)

3. $E_s = E_1 + E_2 + E_3$ (6)

4. $E_s - E_1 - E_2 = 0$ (6)

5. voltage drop (7)

6. counterclockwise (22)

7. negative (15)

8. 3 a. $(I = \dfrac{E}{R})$

9. 21 v. (E = IR)

10. 27 w. $(P = I^2 R)$

11. 180 w. $(P = EI)$

12. clockwise   (49)

13. 40 v.   (53)

14. 2.5 a. $(I = \dfrac{E}{R})$

15. 5 v. $(E = IR)$

16. 12.5 w. $(P = I^2 R)$

17. 2.5 a.   (1)

18. 100 w. $(P = EI)$

19. It decreases.   (67)

20. $R_t = 10\,\Omega$ $\left(\dfrac{1}{R_t} = \dfrac{1}{R_1} + \dfrac{1}{R_2} + \dfrac{1}{R_3}\right)$

21. $I_t = 2$ a. $(I = \dfrac{E}{R})$

22. 0.625 a. $(I = \dfrac{E}{R})$

23. 2.5 w. $(P = \dfrac{E^2}{R})$

24. 40 w. $(P = EI;\ P = I^2 R;\ P = \dfrac{E^2}{R})$

25. $R_t = 1\,\Omega$ (74)

26. $R_t = 10\,\Omega$ (76)

CHAPTER FIVE

Direct-Current
Compound Circuits

In Chapter Four you practiced solving circuit values in series and parallel circuits. However, few circuits in actual use are either pure series or pure parallel circuits; most are a combination of the two. They are called series-parallel, or compound, circuits.

When you finish this chapter you will be able to:

- reduce compound circuits to their simplest form;

- solve for current, voltage, resistance, and power in compound circuits; and,

- solve for currents and voltages in voltage dividers.

Series-Parallel Combinations

1. At least three resistors are required to form a compound circuit. (Remember that a resistor could be any circuit component across which voltage is dropped—a fuse, a lamp, or some other device.) A one-resistor circuit must be series. A two-resistor circuit could be either series or parallel, but not a combination of the two. Why must a circuit containing

only one resistor be a series circuit? _____

_____

- - - - - - - - - -

There is only one path for current flow.

Figure 5-1. R_1 in series with parallel combination of R_2 and R_3. Refer to Figure 5-1 for Frames 2 through 16.

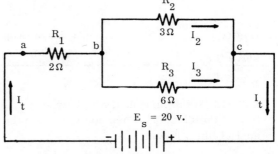

2. To better understand a compound circuit, we identify portions of the circuit as either series or parallel. Look at the compound circuit in Figure 5-1. Would the portion of the circuit between points a and b be considered series or parallel? _____ Why? _____

_____

- - - - - - - - - -

Series, because the total circuit current flows in that portion of the circuit (or words to that effect).

3. The portion of the circuit between points b and c is parallel because

_____.

- - - - - - - - - -

The total circuit current divides into two separate paths in that portion of the circuit (or words to that effect).

4. Is R_t equal to the sum of R_1, R_2, and R_3 in Figure 5-1? _____ Why or why not? _____

- - - - - - - - - -

No, R_t is not the sum of the three individual resistances, because the resistance of the R_2/R_3 combination is not the sum of those resistors, since they are in parallel.

5. What is the combined resistance (R_t) of the resistors R_2 and R_3? _____

- - - - - - - - - -

2 Ω   (Did you remember to use the "product over sum" shortcut?)

6. What is R_t for the complete circuit in Figure 5-1? _____

- - - - - - - - - -

4 Ω   (Add the resistance of R_1, 2Ω, and the combined resistance of R_2 and R_3, 2Ω.)

7. Before you can determine the voltage drop across R_1, what must you solve for? _____

- - - - - - - - - -

I_t (The total current flows through R_1, because that is the series portion of the circuit. Then you will have the values I and R_1 so you can apply Ohm's Law.)

8. $I_t =$ _____

- - - - - - - - -

5 a. $(I_t = \dfrac{E_s}{R_t} = \dfrac{20 \text{ v.}}{4\,\Omega} = 5 \text{ a.})$

9. What is the voltage drop across R_1? _____

- - - - - - - - -

10 v. $(E_{R_1} = I_t R_1 = 5 \text{ a.} \times 2\,\Omega = 10 \text{ v.})$

10. The branch current I_2 is $3\frac{1}{3}$ a. What is the voltage drop across R_2?

_____

- - - - - - - - -

10 v. $(E_{R_2} = I_2 R_2 = 3\frac{1}{3} \text{ a.} \times 3\,\Omega = 10 \text{ v.})$

11. The branch current I_3 is $1\frac{2}{3}$ a. What is the voltage drop across R_3?

_____

- - - - - - - - -

10 v. $(E_{R_3} = I_3 R_3)$

12. Now let's see how adding another resistance in parallel changes all the
values of voltage and current in the circuit shown in Figure 5-1. If we
add another resistor in parallel with R_2 and R_3, the total resistance of
the parallel branches will decrease. (R_1 will not change, of course, because it is a fixed value.) Since R_t of the circuit has decreased (R_1 plus
a lower value of parallel resistance), I_t must increase, because of the
equation $I = \dfrac{E}{R}$. The increased I_t will cause an increased voltage drop
across R_1 because of the equation $E = IR$. To see how this works, we will
add a 10-ohm resistor in parallel with R_2 and R_3. For simplicity (and
because this is standard practice) we will round off to two decimal places
in our calculations.

First find R_t of the parallel branches, which now have resistances of
$3\,\Omega$, $6\,\Omega$, and $10\,\Omega$.

$$\frac{1}{R_t} = \frac{1}{3} + \frac{1}{6} + \frac{1}{10}$$

$$\frac{1}{R_t} = 0.33 + 0.17 + 0.10$$

$$\frac{1}{R_t} = 0.6$$

$$R_t = \frac{1}{0.6}$$

$$R_t = 1.67 \, \Omega$$

R_t for the whole circuit is now 3.67Ω, since R_1 (2Ω) is added to the parallel resistance of 1.67Ω.

$$I_t = \frac{E_s}{R_t} = \frac{20}{3.67} = 5.45 \text{ a.}$$

The voltage drop across R_1 is found by the equation $E = IR$, or 5.45 a. x 2Ω = 10.9 v.

By Kirchhoff's Law of Voltages, the voltage drop across the parallel branches is 20 v. − 10.9 v. (the source voltage minus the voltage drop across R_1), or 9.1 v.

Thus, the voltage drop across the parallel portion of a circuit depends on the voltage drops in the series portion, but the voltage drop across each branch of the parallel portion is the same. A parallel portion of the circuit can be regarded as a single resistor whose value is the R_t of all the branches.

When a resistance is added in parallel with a parallel portion of a circuit, does the total resistance (R_t) of that portion increase or decrease?

_____ Why ? _____

_____

- - - - - - - - -

It decreases, because of the equation $\frac{1}{R_t} = \frac{1}{R_1} + \frac{1}{R_2} + \frac{1}{R_3}$

(If you said that it decreases because another path for current flow was provided, you are also correct and probably understand parallel resistance even better.)

13. If the circuit shown in Figure 5-1 is changed as described in Frame 12, does the voltage drop across R_1 increase or decrease? _____

Why ? _____

- - - - - - - - -

It increases, because of Kirchhoff's Law of Voltages: $E_s − E_1 − E_2 = 0$. Less voltage is dropped across the parallel resistors, so more is dropped across R_1.

14. The voltage drop across one parallel branch is the same as the voltage drop across all other branches. However, the current through each branch depends on the resistance of that branch. (Remember that resistors are physical devices whose values are fixed.) When the resistance of any branch is changed, or when another branch is added, the current (I_t) flowing in the series portion of the circuit is divided among the branches according to the individual resistance of each branch. I_t flows into the parallel portion, and I_t flows out, in accordance with Kirchhoff's Current Law. Circle the Ohm's Law equation used to find the current of any branch of a parallel circuit.

$$E = IR \qquad I = \frac{E}{R} \qquad R = \frac{E}{I}$$

- - - - - - - - - -

$I = \dfrac{E}{R}$

15. You must keep in mind that all the source voltage in a series circuit must be accounted for in individual voltage drops around the circuit. A parallel portion of a series-parallel circuit can be regarded as a single resistor whose value R_t is the effective resistance of that portion. Can you make a general statement about the voltage drops across the branches of a parallel circuit, or across the parallel portion of a series-parallel circuit?

_____

_____

- - - - - - - - - -

The voltage drop across any branch of a parallel circuit is the same as the voltage drop across any other branch.

16. You can be sure that branch circuits are truly parallel if they have a common junction at either end. A common junction is either a point at which the total circuit current I_t divides into the branch currents or a point at which the branch currents rejoin to form the total current. One common junction for R_2 and R_3 is point b in Figure 5-1. (Point b is also common to one end of R_1, but we are concerned with the parallel branches.) What is another common junction point? _____

- - - - - - - - - -

point c

17. Remember that the same difference in potential exists across all branches of a parallel circuit. However, the total current in the series-parallel circuit depends on the effective resistance of the parallel portion and on the other resistances in series with it. Now let's solve some problems in another series-parallel circuit.

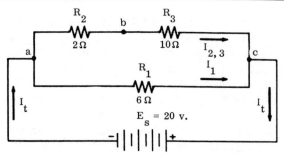

Figure 5-2. R_1 in parallel with the
series combination of R_2 and R_3.

Refer to Figure 5-2 for Frames 17 through 22.

Figure 5-2 shows a circuit in which two resistors <u>in series</u> (R_2 and R_3) form one branch of a parallel circuit. The total current I_t flows <u>into</u> the parallel circuit, splitting into two branches at point a. At what point do the branch currents <u>rejoin</u> to form I_t? _____

- - - - - - - - - -

point c

18. The source voltage E_s (20 v.) is dropped between points a and c. The largest voltage drop is across which resistor? _____ Why? _____

_____

- - - - - - - - - - -

R_1, because the 20 v. must be divided between R_2 and R_3.

19. The next few frames will show you a systematic way to find the branch currents. What is the resistance of the top branch of the circuit in Figure 5-2? _____

- - - - - - - - - -

12 Ω   (Add R_2 and R_3.)

20. R_t = _____

- - - - - - - - - -

4 Ω   (The "Product over sum" solution is $R_t = \dfrac{6 \times 12}{6 + 12} = \dfrac{72}{18} = 4.$)

21. What is the current through the top branch ($I_{2,3}$)? _____

- - - - - - - - - -

1. 67 a. $(I = \dfrac{E}{R} = \dfrac{20 \text{ v.}}{12\,\Omega} = 1.67$ a.

22. $P_t =$ _____ (You do not need to solve for I_t first.)

- - - - - - - - - - -

100 w.   (Since you already knew E_s and R_t, you should have used the equation $P_t = E_s^2/R_t$.)

Refer to Figure 5-3 for Frames 23 through 27.

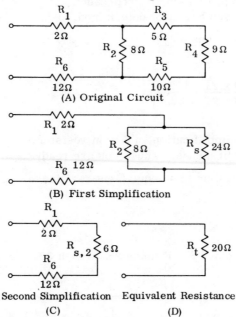

(A) Original Circuit

(B) First Simplification

Second Simplification   Equivalent Resistance
(C)   (D)

Figure 5-3. Solving total resistance
in a compound circuit.

23. Compound circuits may be made up of any number of resistors arranged in numerous series and parallel combinations. Before you can arrive at R_t, the equivalent resistance of the entire circuit, you must first reduce the circuit to its series and parallel equivalents. The original circuit in Figure 5-3 has six resistors. Name the four resistors that make up the parallel portion of the circuit. _____

- - - - - - - - - - -

R_2, R_3, R_4, and R_5

24. The parallel portion of the circuit has how many branches? _____

- - - - - - - - - - -

two

25. In the original circuit of Figure 5-3, one branch of the parallel portion has a single resistor, R_2, whose resistance is 8Ω. The other branch has three resistors in series. What is their combined resistance? _____

- - - - - - - - - -

24Ω

26. In the first simplification, R_S represents the <u>series</u> equivalent of R_3, R_4, and R_5, which you have just determined. What is the total resistance of the <u>parallel branches only</u>? _____

- - - - - - - - - -

6Ω $(R_t = \dfrac{8 \times 24}{8 + 24} = \dfrac{192}{32} = 6)$

27. In the second simplification, a 6-ohm resistor represents the resistance of the parallel branches. Thus, you are ready to reduce the series-parallel combination to an equivalent resistance, R_t, of _____

- - - - - - - - - -

20Ω

Refer to Figure 5-4 for Frames 28 through 32.

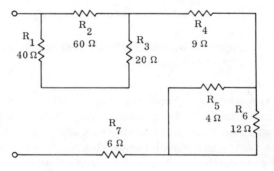

Figure 5-4. Series-parallel combination of resistors.

28. Now let's try a different kind of compound circuit. Figure 5-4 shows a circuit that has parallel portions alternating with single resistors. This circuit is more complicated, so it is important to be sure you can trace through the circuit to identify the series and parallel portions. The

placement of the resistors can be misleading, but you will stay on the right track if you remember exactly what constitutes a parallel portion. Think of the circuit in terms of current flow. The source voltage is not shown, so you have no way of knowing the direction of current flow. However, in this case it doesn't matter. Start at one of the terminals and trace around the circuit. Each time you find a point where the current (if there were a voltage source) would have to split into branches, you know you have one end of a parallel portion of the circuit. (It might help to put a large dot at that point.) When you get to a point where those same branch currents flow together again, you have the other end. Don't be confused by the number of resistors. Any parallel branch could have several resistors in series. You know they are in series if the same current must flow through each. After you have identified all the parallel portions of the circuit, the series portion of the entire circuit includes any resistor through which current flow is the same as the current flow entering the circuit from the voltage source, that is, the total current I_t. In Figure 5-4, name the resistors that make up the series portion of the entire circuit; that is, the resistors that are not found in any parallel branch.

_____

– – – – – – – – –

R_4 and R_7

29. The circuit has two separate parallel combinations. Resistors R_5 and R_6 form one parallel combination. Name the resistors that form the other.

_____

– – – – – – – – – –

R_1, R_2, and R_3

30. The equivalent resistance of R_1, R_2, and R_3 is _____.

– – – – – – – – – –

$30\,\Omega$

31. The equivalent resistance of R_5 and R_6 is _____.

– – – – – – – – – –

3Ω

32. The total circuit resistance, R_t, is _____.

– – – – – – – – – –

48Ω

Refer to Figure 5-5 for Frames 33 through 51.

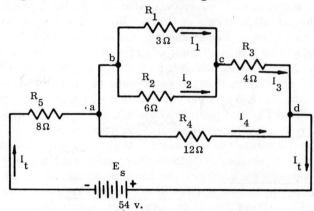

Figure 5-5. A compound circuit.

33. Figure 5-5 shows an even more complicated circuit, because it includes a parallel portion <u>within</u> a larger parallel portion. You have already observed that subscripts are used for labeling both circuit components and their associated values. (However, a number or letter that is a subscript in text might be hand-lettered that way on a schematic diagram.) Any convenient method of labeling may be used as long as it is descriptive. The voltage drop across R_1, for example, could be designated E_{R_1} or E_{bc}. Either correctly identifies the points between which the voltage would be measured. The voltage drop across R_4 could be designated E_{R_4} or _____.

- - - - - - - - - -

E_{ad}

34. It is often necessary to make intermediate calculations before you can arrive at a specific unknown. For example, before you can solve for I_1, you must first know the voltage drop across R_1 (E_{bc}). Before you can solve for I_t, you must know _____.

- - - - - - - - - -

R_t

35. Computations can become very confusing unless you keep track of values as you solve them. It is a good idea to write in values on the circuit diagram or to write them on a separate piece of paper. The advantage of writing them on the diagram is that it makes any electrical situation easier to visualize. Before you can find R_t, you must simplify the circuit in successive steps. R_1 and R_2 are parallel branches between points b and c, so their equivalent resistance can be designated R_{bc}. But R_{bc} is part

of one of the parallel branches connected between points _____ and

_____.

- - - - - - - - - -

à and d

36. R_{bc} = _____

- - - - - - - - - -

$2\,\Omega$

37. R_{bd} includes what three resistors? _____

- - - - - - - - - -

R_1, R_2, and R_3

38. R_{bd} = _____

- - - - - - - - - -

$6\,\Omega$

39. R_{ad} = _____

- - - - - - - - - -

$4\,\Omega$

40. You have now arrived at an equivalent circuit that includes the battery E_s, R_{ad}, and R_5. Does the circuit need to be further simplified? Why?

_____

- - - - - - - - - -

No, because it is now an equivalent series circuit.

41. R_t = _____

- - - - - - - - - -

$12\,\Omega$

42. By Ohm's Law, the line current (another term for I_t) is _____.

- - - - - - - - - -

4. 5 a.

43. The line current flows through R_5, so the voltage drop E_5 is _____.

- - - - - - - - - -

   36 v.

44. According to Kirchhoff's Law of Voltages, the sum of the voltage drops
    around the circuit is equal to the source voltage. Therefore, E_{ad} is

   _____.

- - - - - - - - - -

   18 v.

45. $I_4 = $ _____   (Hint: You know E_{ad} and R_4.)

- - - - - - - - - -

   1. 5 a.

46. There are only two parallel branches between points a and d. One of
    these branches, however, includes another pair of parallel branches.
    You already know that I_t is 4.5 a. and I_4 is 1.5 a. Therefore, the cur-

    rent flowing into the junction at point b is _____.

- - - - - - - - - -

   3 a.   ($I_t - I_4$, or 4.5 a. − 1.5 a.)

47. According to Kirchhoff's Current Law, all the current flowing into the

    junction at point b flows out at point c. $I_3 = $ _____

- - - - - - - - - -

   3 a.

48. E_3 (the voltage drop across R_3) is _____.

- - - - - - - - - -

   12 v.

49. Since E_{ad} is 18 v. and E_3 is 12 v. , what is E_{bc}? _____

- - - - - - - - - -

   6 v.

50. $I_1 = $ _____

- - - - - - - - - -

2 a.

51. $I_2 =$ _____

- - - - - - - - - -

1 a.   (Note: You could have solved I_2 either by the equation $I_2 = \dfrac{E_2}{R_2}$ or by $I_2 = I_3 - I_1$.)

52. You have been able to use the "product over sum" shortcut in solving parallel resistance in the last several frames, but remember that you must sometimes use the regular equation, which involves reciprocals:
$\dfrac{1}{R_t} = \dfrac{1}{R_1} + \dfrac{1}{R_2} + \dfrac{1}{R_3} \ldots$ (etc.). Solve R_t in the partial circuit diagram below. $R_t =$ _____

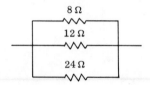

$8\,\Omega$

$12\,\Omega$

$24\,\Omega$

- - - - - - - - - -

$4\,\Omega$ (Note: If you use fractions, the answer comes out even. If you carry to two decimal places, the answer is $4.03\,\Omega$.)

53. Can you think of another way to solve for R_t that avoids reciprocals?

_____

_____

- - - - - - - - - -

Perhaps not; but there is a way. You could have solved for the equivalent resistance of two branches using the "product over sum" shortcut. Then you could have repeated the shortcut with that equivalent resistance and the third branch. Try it. Your answer should be the same, $4\,\Omega$.

54. As you have learned in this section, any series-parallel circuit can be reduced to an equivalent series circuit. This equivalent circuit is useful to find R_t and I_t, as well as to determine the difference in potential across any parallel portion of a compound circuit. I_t splits into at least two paths when it enters a parallel portion; the voltage drop across that portion (which is the same across all branches) can be used to find each branch current, simply by dividing the branch resistance into the voltage across that branch.

Now let's put it all together by solving some problems in one more compound circuit. The draftsmen who draw schematic diagrams do not like to waste space or lines, so the parallel portions of a circuit are often far from obvious. Such a circuit appears below. It might be helpful to redraw the circuit to make the parallel portions more clear, or to place dots at the end junctions of parallel branches. You are asked to solve for only a few values; however, you will have to find other values before you have enough information to find the specific values requested. If you feel you need more practice, solve for some additional circuit values, such as other voltage drops and branch currents. You can check your own answers, because all values will be consistent (if they are correct) with the values given in the answer to this frame. (For example, the sum of branch currents for each parallel portion must be equal to I_t.)

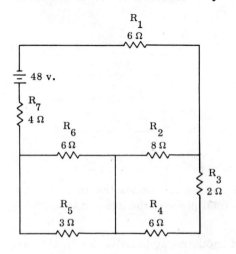

(a) What is the voltage drop across R_1?
(b) What is the voltage drop across R_4?
(c) What is the current through R_6?

- - - - - - - - - -

If you redrew the circuit, it might look something like this:

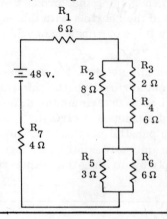

(a) The voltage drop across R_1 is 18 v.

    1. Find R_t of each parallel portion: R_t of R_2, R_3, and R_4 is $4\,\Omega$; R_t of R_5 and R_6 is $2\,\Omega$. R_t of entire circuit is $16\,\Omega$ $(6\,\Omega + 4\,\Omega + 2\,\Omega + 4\,\Omega)$.

    2. $I_t = \dfrac{E_s}{R_t} = \dfrac{48 \text{ v.}}{16\,\Omega} = 3$ a.

    3. $E_{R_1} = I_t R_1 = 3$ a. $\times 6\,\Omega = 18$ v.

(b) The voltage drop across R_4 is 9 v.

    1. First find the voltage drop across this parallel portion. $E = I_t R = 3$ a. $\times 4\,\Omega = 12$ v.

    2. Find current through R_4. $I_{R_4} = \dfrac{12 \text{ v.}}{8\,\Omega} = 1.5$ a.

    3. $E_{R_4} = 1.5$ a. $\times 6\,\Omega = 9$ v.

(c) The current through R_6 is 1 a.

    1. Find the voltage drop across this parallel portion. $E = I_t R = 3$ a. $\times 2\,\Omega = 6$ v.

    2. $I_{R_6} = \dfrac{E}{R_6} = \dfrac{6 \text{ v.}}{6\,\Omega} = 1$ a.

Voltage Dividers

55. In practically all electronic devices, such as radio receivers and transmitters, certain design requirements (which require different voltages) recur many times. It is both impractical and unnecessary to have a separate power supply for each voltage requirement, because the same result can be achieved by <u>voltage dividers</u>.

To understand the discussion that follows, you need to know a few new terms and a new symbol. A ground is merely a common connection point—usually the metal chassis of the equipment. For example, several wires might be connected to the sheet metal frame of a radio. All such connection points are electrically the same, so each point in the schematic could be represented by the "ground" symbol. If one side of the power supply is connected to ground, every point connected to ground is also electrically connected to that side of the power supply. Normally the grounded side is negative. In the diagram below, the negative side of the battery and R_1 are both grounded.

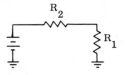

Draw the symbol for "ground."

- - - - - - - -

⏚ (Note: The number of horizontal lines is not important, but each
line is shorter than the one above it. The lowest line could be a dot.)

56. Many devices draw very small amounts of current. To reduce the current
to the value required, large values of resistance must be used. Two common prefixes in electricity are "milli-," which means 1/1000th (or 0.001)
and "kilo-," which means 1,000. One milliampere is written "1 ma."
One kilohm is usually abbreviated 1 kΩ or simply 1 K. (Don't be confused
by the interchangeability of small and capital letters in some abbreviations; conventions differ somewhat.)

   (a) If the source voltage is 100 v. and the total circuit resistance is

      50 K, the current is 0.002 a., or _____ ma.

   (b) If the source voltage is 40 v. and the total circuit resistance is

      10 K, the current is _____ ma.

   (c) The value 6 ma. is the same as _____ a.

   (d) 50 K is the same as _____ Ω.

- - - - - - - - - -

(a) 2; (b) 4; (c) 0.006; (d) 50,000. 50 K could also be written 50 kΩ.
(Note: For help in working with very large numbers and small decimals,
you may wish to review the laws of exponents in Appendix III, page 278.)

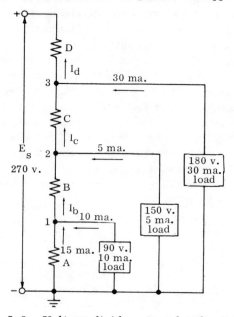

Figure 5-6. Voltage divider, to solve for R and P.

Refer to Figure 5-6 for Frames 57 through 66.

57. A typical voltage divider consists of two or more resistors connected in series across the primary power supply. The primary voltage E_S must be as high as or higher than any of the individual voltages it is to supply. As the primary voltage is dropped by successive steps through the series resistors, any desired fraction of the original voltage may be "tapped off" to supply individual requirements. The value of each series resistor used is determined by the voltage to be dropped across it. In Figure 5-6, a voltage divider is connected across a 270-volt source E_S. (The voltage source is often indicated on schematics by showing only the terminals.) The voltage divider in Figure 5-6 consists of four resistors, labeled A, B, C, and D. The points (1, 2, and 3) at which partial voltages are tapped off are called "taps." In this case, external loads are connected to all three taps, and the voltage and current requirements of each load are indicated on the schematic. The load placed across resistor A is 90 v. and 10 ma. The voltage drop across A, measured between ground and tap 1,

is _____ v.

- - - - - - - - - -

90

58. The Ohm's Law equations have so far been based on the basic units: ohms, amperes, and volts. But now we are starting to use milliamperes (called "milliamps") and kilohms. These should cause no problem as long as you are careful to use the appropriate decimals for milliamperes and the correct number of zeroes for kilohms. But since these two units of measurement are quite common, you should know a short cut. Since $E = IR$, you know that 1 a. flowing through a 1-ohm resistor drops 1 v. A current of

1 ma. flowing through a 1-kilohm resistor drops _____ v.

- - - - - - - - - -

1   ($E = 0.001 \times 1,000 = 1$)

59. Thus, in an equation where all resistances are in kilohms and all currents are in milliamperes, voltage is in _____.

- - - - - - - - - -

volts

60. According to Kirchhoff's Law of Currents, the current flowing through resistor B must be the sum of the current flowing through A and that flowing through the load across A. I_b is _____.

- - - - - - - - - -

25 ma.

61. The required voltage for the 150-volt load is obtained by selecting a tap on the voltage divider at which the potential difference (between the tap and ground) is 150 v. The combined voltage drop across resistor A and resistor B is also 150 v. Why? _____

    _____

    - - - - - - - - -

    Because the voltage drop across all parallel branches is the same.

62. The current I_c is 30 ma. How do we know? _____

    _____

    - - - - - - - - - -

    The load current of 5 ma. is added to current I_b, which we know is 25 ma.

63. What is current I_d? _____

    - - - - - - - - - -

    60 ma.   (I_c plus the load current of 30 ma.)

64. I_t in Figure 5-6 flows through only one resistor of the voltage divider. Which one? _____

    - - - - - - - - - -

    D.   (This resistor is not part of any parallel portion of the circuit. It is in series with the voltage source.)

65. The correct voltage requirements are supplied to the three loads in Figure 5-6 because of the varying differences in potential at taps 1, 2, and 3. These differences in potential are with respect to ground. For example, the voltage drop across resistor A is 90 v. This is also the difference in potential between tap 1 and ground. The voltage drop across resistor B is 60 volts. The difference in potential, with respect to ground, between tap 2 and ground is 150 v. Thus, the two voltage drops of 90 v. and 60 v. represent a difference in potential of 150 v. The voltage drop across resistor C (between taps 2 and 3) is 30 v. What is the difference in potential between tap 3 and ground? _____

    - - - - - - - - - -

    180 v.

66. We now know all the values of voltage and current necessary to find the values of resistance in the circuit. For example, we can find the value of resistor A because we know the current through it and the voltage dropped across it.

   (a) $R_A =$ _____

   (b) $R_B =$ _____ (Hint: First find E_B. You already know E_A and E_{A+B}.)

   (c) $R_C =$ _____

   (d) $R_D =$ _____

- - - - - - - - - -

(a) 6 K $(R_A = \dfrac{E_A}{I_A} = \dfrac{90 \text{ v.}}{15 \text{ a.}} = 6 \text{ K})$

(b) 2.4 K $(E_B = 150 \text{ v.} - 90 \text{ v.} = 60 \text{ v.})$

(c) 1 K $(E_C = 180 \text{ v.} - 150 \text{ v.} = 30 \text{ v.})$

(d) 1.5 K

67. We have seen, in working the problems associated with Figure 5-6, that the voltage divider is an excellent application of the properties of parallel circuits. Each external device, or "load," connected to one of the taps on the voltage divider is a branch in parallel with <u>part</u> of the voltage divider. As we move to a tap farther away from ground, which is the negative end of the voltage source in Figure 5-6, we select a larger difference in potential between that tap and ground. Since all branches of a parallel circuit "see" the same difference in potential, we can select a voltage appropriate to the external load. Remember that adding another resistance in parallel with part of the circuit will <u>decrease the resistance</u> of that part of the circuit. This will in turn decrease the difference in potential across the parallel circuit. Thus, the voltage across the parallel circuit is lower, but the voltage across the <u>series</u> resistors is correspondingly greater. This is consistent with Kirchhoff's Law of Voltages. It makes sense, because the total current flowing in the circuit has also increased (since the total resistance is lower).

   The circuit shown on the following page shows a voltage divider consisting of three one-kilohm resistors connected to a 300-volt source. <u>Before</u> the load is connected, the total circuit current is 100 ma. (0.1 a.) and each resistor sees a voltage drop of 100 v. The dashed line indicates that the two-kilohm load has not been connected to the tap.

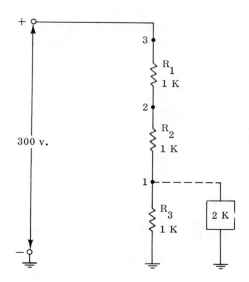

After the load is connected:

(a) What is the difference in potential between tap 3 and ground?

_____

(b) What is the resistance between tap 1 and ground? _____

(c) What is the current through R_2? _____

(d) What is the difference in potential between tap 1 and ground?

_____

(e) What is the current through the two-kilohm load? _____

- - - - - - - - - -

(a) 300 v.   (Same as the source voltage.)
(b) 0.67 K   (Use the "product over sum" method.)
(c) 112.36 ma.   (E_s divided by R_t, or 300 v. ÷ 2.67 K.   This is I_t.)
(d) 75.28 v.   (I_t times 0.67 K)
(e) 37.64 ma.   (75.28 v. ÷ 2 K)

68. Now let's see how currents are distributed when external loads are connected to a voltage divider. In Figure 5-7, on the following page, resistances are given for R_4, R_5, R_6, and R_7 (the voltage divider), but not for
the three loads, R_1, R_2, and R_3. The source voltage is known
(E_s = 510 v.), and enough information about currents is given to enable
us to work out all the voltage drops in the circuit; but the only way to go
about it is to apply Kirchhoff's Law of Currents.

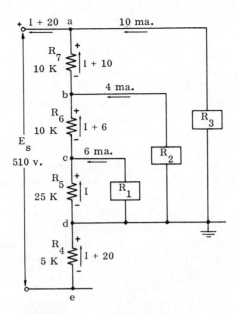

Figure 5-7. Voltage divider, to
solve for E and R.

Refer to Figure 5-7 for Frames 68 through 82.

The only resistor through which <u>all</u> circuit current flows is _____.

- - - - - - - - - -

R_4

69. The load currents, as shown in the diagram, are 6 ma., 4 ma., and 10 ma.
The only resistor through which <u>no</u> external load current flows is _____.

- - - - - - - - - -

R_5

70. Thus, if we first solve for I (the current through R_5), we can easily find
the current through the other resistors of the voltage divider (R_6 and R_7).
The current through R_6 is I plus 6 ma. (the current through R_1); and the
current through R_7 is I plus 6 ma. plus 4 ma. (the current through R_2).
Naturally, once we know both the current and the resistance for any resistor, we can find the voltage drop across that resistor. We shall also

know the voltage across each external load. Why? _____

_____

- - - - - - - - - -

Each load is in parallel with a portion of the circuit for which the voltage drop can be found, and we know that voltage drops across parallel branches are equal.

71. To find I, the current through R_5—since we don't know the voltage drop across R_5—we must set up an equation that expresses all voltage drops in terms of current and resistance ($E = IR$). Start with your knowledge that all voltage drops in the circuit must equal E_s:

$$E_4 + E_5 + E_6 + E_7 = E_s$$

You know that the current through R_5 (I) and all load currents (20 ma. in all) flow through R_4. You can express E_4, then, in terms of current and resistance: $E_4 = 5(I + 20)$. I is the current that does not flow in the loads, and 20 ma. represents the total load current. The resistance of R_4 is 5 K. You are working with milliamperes and kilohms, so the resistance is simply 5 in the equation. Complete the equation that will express all voltage drops, expressing each voltage drop in terms of R and I.

$$5(I + 20) + 25I + \underline{\hspace{1cm}} + \underline{\hspace{1cm}} = 510$$

- - - - - - - - - -

$5(I + 20) + 25I + 10(I + 6) + 10(I + 10) = 510$

72. Now collect your terms, and your equation is simplified as:

$$5I + 100 + 25I + 10I + 60 + 10I + 100 = 510$$

$$50I + 260 = 510$$

$$50I = 250$$

$$I = \underline{\hspace{1cm}}$$

- - - - - - - - - -

5 ma.

73. The current through R_5 is _____.

- - - - - - - - - -

5 ma.

74. The current through R_4 is _____.

- - - - - - - - - -

25 ma.

75. $E_4 = $ _____

- - - - - - - - - -

125 v.

76. $E_5 = $ _____

- - - - - - - - - -

125 v.

77. What is the voltage drop across R_1? _____

- - - - - - - - - -

125 v.   (R_1 is in parallel with R_5.)

78. What is the resistance of R_1? _____

- - - - - - - - - -

20.83 K

79. What is the voltage across R_2? (Hint: First you have to find E_6.) _____

- - - - - - - - - -

235 v.   ($E_5 + E_6$, or 125 v. + 110 v., since R_2 is in parallel with R_5 and R_6.)

80. How much power is absorbed by R_3? _____

- - - - - - - - - -

3.85 w.   (The voltage drop across R_3 is $E_5 + E_6 + E_7$, or 385 v. $P = EI = 385 \times 0.01 = 3.85$.)

81. How much power is absorbed by R_2? _____

- - - - - - - - - -

0.94 w.

82. $P_t = $ _____

- - - - - - - - - -

12.75 w.   ($P_t = E_s I_t = 510$ v. $\times$ 25 ma. = 12.75 w.)

The circuits described in this chapter are the kind most frequently en-countered in actual practice; that is, they have both series and parallel elements. It is necessary to look at the circuit as a whole to find some of the values. For example, you need to know I_t before you can solve for the voltage across parallel branches—and this voltage is needed to solve for the current through a specific branch.

You must approach the direct-current compound circuit systematically, first identifying the parallel portions and then reducing the entire circuit to its simplest form: an equivalent series circuit. Once you have done this, it is not difficult to solve any problem of current, voltage, resistance, or power.

The voltage divider is merely one application of a compound circuit. Each external load tapped into the voltage divider is merely a branch of a parallel circuit. However, its resistance changes the equivalent resistance of one portion of the circuit, and thus, the values of voltage and current throughout the circuit.

If you feel you understand the basic concepts of the compound circuit and the voltage divider, go on to the Self-Test.

Self-Test

These questions will test your understanding of Chapter Five. Write your answers on a separate sheet of paper and check them with the answers provided following the Self-Test.

Refer to the diagram below for questions 1 through 4.

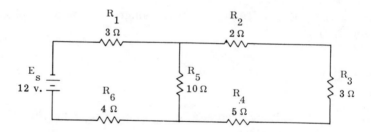

1. What is the total resistance R_t of the circuit?

2. What is the total current I_t in the circuit?

3. What is the voltage across R_3?

4. What is the total power consumed in the circuit?

5. What is R_t, the equivalent resistance, of the circuit below?

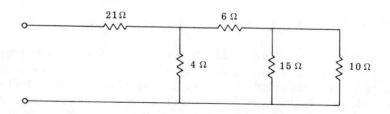

6. Refer to the diagram below.

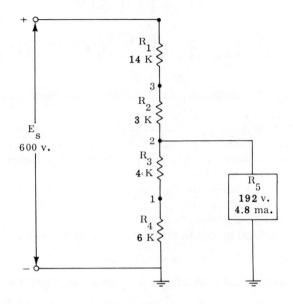

(a) What is the current through R_1 ?
(b) What is the voltage across R_4?
(c) What is the difference in potential between tap 3 and ground?
(d) How much current would flow in the circuit if R_5 were
disconnected?

Answers

If your answers to the test questions do not agree with the ones given below,
review the frames indicated in parentheses after each answer before you go
on to the next chapter. (Where there is no one specific reference, the appropriate equation is given.)

1. 12Ω (6)

2. 1 a.   (8)

3. 1.5 v.   (17−21)

4. 12 w.   (22)

5. 24Ω (23−27)

6. (a) 24 ma.   (67)
   (b) 115.2 v.   (67)
   (c) 264 v.   (67)
   (d) 22.22 ma. $\left(I_t = \dfrac{E_s}{R_t} \right)$

CHAPTER SIX

Magnetism and Electromagnets

The principles of magnetism and electricity are interrelated. Electromagnets are used in some direct-current circuits, and it is impossible to understand alternating-current theory (which will be introduced in Chapter Seven) without some basic understanding of magnetism.

The principles of magnetism go far beyond the treatment given in this chapter; but what you learn here will prepare you to study some of the material presented later in the book.

When you have finished this chapter, you will be able to:

- describe the behavior of magnetic fields;

- distinguish among natural and permanent magnets and electromagnets;

- relate electric current to its associated magnetic field;

- predict the action of a magnetic field about a current-carrying coil;

- describe the major properties of magnetic materials;

- describe the nature and operation of electromagnets; and,

- explain how magnetic shielding is accomplished.

1. A substance is said to be a magnet if it has the power to attract such substances as iron, nickel, or cobalt, known as magnetic materials. Some materials are much more magnetic than others. Aluminum, for example, is not noticeably magnetic, while steel, which is mostly iron, is very magnetic. Non-metals are not magnetic at all.

    A magnet has two <u>magnetic poles</u>, or points of maximum attraction. If you suspend a bar magnet from a string so that it is balanced parallel to the earth and then set it to spinning, it will eventually come to rest with one end pointing north. If you then nudge that end away, it will return to the north-pointing direction. This end of the magnet is called the <u>south pole</u>. Opposites attract in magnetism, too. Knowing that, what do you think the other end is called? _____.

    - - - - - - - - - -

    north pole

2. This principle of the north-seeking pole led to the invention of the compass. The compass needle is a magnetized strip of metal that points to
the north. The end of the needle that indicates direction is the

(north/south) _____ pole.

- - - - - - - - - -

south

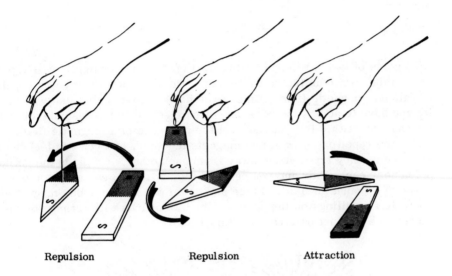

Repulsion   Repulsion   Attraction

Figure 6-1. Laws of attraction and repulsion.

3. Figure 6-1 illustrates that magnets, like charged bodies, follow laws of

attraction and repulsion. A north pole will (attract/repel) _____
another north pole.

- - - - - - - - -

repel

4. What is the equivalent law about the attraction and repulsion of charged

bodies? _____

- - - - - - - - -

Something like, "Like charges repel; unlike charges attract."

5. Can you make a similar statement about magnetic poles? _____

_____

- - - - - - - - - -

Like poles repel; unlike poles attract.

6. Two bar magnets are "stuck" end to end by magnetic force. The end of
one magnet is a north pole. The end of the other is a _____.

- - - - - - - - - -

south pole

7. Like poles of two bar magnets (two north poles, for example) will repel
each other, while unlike poles attract. The field of attraction or repulsion
is called a <u>magnetic field</u>. Unless you have an instrument that will sense
it, you have to accept the presence of an <u>electric</u> field, described in Chapter One, on faith. But you can feel the presence of a magnetic field.
Bring the repelling ends of two magnets together, and you can feel the resistance as they try to move away from each other. Or place the unlike
poles near each other and you can feel the pull. You can also see evidence of a magnetic field. Place a sheet of paper over a bar magnet and
sprinkle iron filings on the paper. The bits of iron will arrange themselves in a distinct pattern, like this:

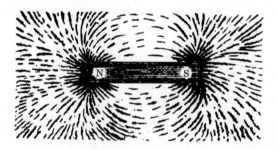

This patterned scattering of iron filings presents a visible "map" of
the _____.

- - - - - - - - - -

magnetic field

8. The magnetic field can be thought of as the lines, known as <u>flux lines</u>, or
<u>magnetic flux</u>, along which a <u>magnetic force</u> acts. By convention, we say
the lines emanate from the north pole of the magnet and re-enter through
the south pole, returning to the north pole through the magnet itself. Figure 6-2 on the next page illustrates the travel of the flux lines outside the

magnet and also shows how these lines of force behave when they are attracted or repelled by another magnet.

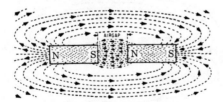

(A) Unlike Poles Attract

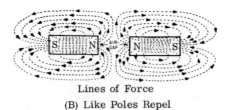

Lines of Force
(B) Like Poles Repel

Figure 6-2. Flux lines between
like and unlike poles.

The outer ends of the magnets are areas in which the flux lines are not acted upon by another magnet. On a separate sheet of paper, draw a single bar magnet. Label the poles and draw some flux lines to indicate their paths outside the magnet. Use arrows to indicate the direction of the flux lines.

- - - - - - - - - -

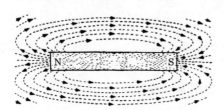

9. If a bar magnet is dipped into iron filings, many of the filings are attracted to the ends of the magnet, but none are attracted to the center. Explain why. _____

- - - - - - - - - -

The points of greatest attraction are the poles of the magnet.

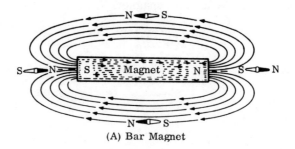

(A) Bar Magnet

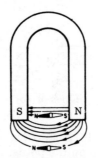

(B) Horseshoe Magnet

Figure 6-3.   Magnetic lines of force.

10. By using a compass, you can observe the presence of the magnetic lines
    of force (flux lines) at various points near the magnet.   Figure 6-3 shows
    the position of the compass needle in various locations.
       As the compass is moved around the magnet, what happens to the com-

    pass needle? _____

    _____

    - - - - - - - - - -

    It swings to keep itself aligned with the flux lines.

11. Magnets may be conveniently divided into three groups: natural magnets,
    permanent magnets, and electromagnets.   You can probably distinguish
    among the three groups from the following descriptions.   Fill in the blank
    following each description with one word: "natural," "permanent," or
    "electromagnet."
       (a) An iron compound called magnetite, found in nature, has been

       known for centuries to attract iron.   _____
       (b) When an electric current is passed through a coil around an iron

       core, a magnet is produced as long as current flows.   _____

(c) Commercial magnets are made by a special process that mag-
netizes certain steels or other alloys which then retain their magnetic properties almost indefinitely. _____

- - - - - - - - - -

(a) natural; (b) electromagnet; (c) permanent

12. "Loadstones" (or "lodestones"), for centuries the only kind of magnet
known, are _____ magnets.

- - - - - - - - - -

natural

13. The magnet in your telephone receiver is produced by a special process
that gives the metal long-lasting magnetic properties. It is called a

_____ magnet.

- - - - - - - - - -

permanent

14. An electromagnet is a (permanent/temporary) _____
magnet.

- - - - - - - - - -

temporary

15. The nature of a magnetic field, which is made up of all the magnetic lines
of force, is important because it has many applications in electricity.
One might think of the lines of force, which form closed loops in space
and through the magnet, as a magnetic circuit. All the lines of force together are called magnetic flux. Flux in a magnetic circuit compares to

what in an electric circuit? _____

- - - - - - - - - -

the electric current

16. Some magnetic fields are, of course, stronger than others. The strength
of the field is related to <u>flux density</u>. The number of lines of force per
unit area is directly porportional to this field strength, or flux density. If
the current flowing through an electromagnet is increased to make the mag-

net stronger, what happens to the flux density? _____

- - - - - - - - - -

It increases.

17. Let's review some terminology:
        (a) The space surrounding a magnet, in which the magnetic force

        acts, is called the _____.
        (b) The entire quantity of magnetic lines of force surrounding a mag-

        net is called _____.
        (c) The number of lines of force per unit area is called _____

        _____.

    - - - - - - - - - -

        (a) magnetic field; (b) magnetic flux, or simply, flux; (c) flux density

Electromagnetism

18. You have learned that magnets are divided into three categories: natural
     magnets, permanent magnets, and electromagnets. Electromagnetism is
     important to the study of electricity because many devices, such as circuit
     breakers and relays, make use of electromagnets.
        Electromagnetism includes two closely related areas: magnetism as it
     is affected by electric current flow, and electricity as it is affected by
     magnetism. You have probably noticed that your automobile radio is af-
     fected when you drive under high-tension power lines. This is an example
     of electricity (the performance of your radio) affected by the magnetic
     field around the power lines. If your home uses circuit breakers instead
     of fuses, they operate when there is an overload of current because the
     current produces magnetism to "trip" the circuit breaker. Both these

     phenomena are examples of _____.

    - - - - - - - - - -

     electromagnetism

19. Magnetism and electricity are so closely related that one cannot be stud-
     ied at length without involving the other. It is important to remember one
     general relationship between magnetism and electricity: Electric current
     flow will <u>always</u> produce some form of magnetism. You can produce a
     weak magnetic field around a flashlight, for example, simply by doing

     what ? _____

    - - - - - - - - - -

     turning it on

20. Electric current flow will (sometimes/always/never) _____ produce some form of magnetism.

- - - - - - - - -

always

Magnetic Field Around a Current-Carrying Conductor

21. A definite relation between magnetism and electricity was discovered in 1819 when a Danish physicist, Hans Christian Oersted, established that an electric current is accompanied by certain magnetic effects and that these effects obey definite laws. If a compass is placed in the vicinity of a current-carrying conductor, the needle aligns itself at right angles to the conductor, thus indicating the presence of a _____.

- - - - - - - - - -

magnetic force or magnetic field

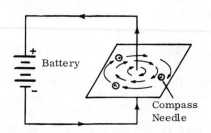

Figure 6-4. Magnetic field around a current-carrying conductor.

22. The presence of a magnetic force can be demonstrated by passing an electric current through a vertical conductor which passes through a horizontal piece of cardboard, as illustrated in Figure 6-4. The presence of a magnetic force is indicated by the positioning of a compass at various points on the cardboard and noting the deflection of the needle. In Figure 6-4, the direction of the magnetic field, as you view the cardboard from

above, is (clockwise/counterclockwise) _____.

- - - - - - - - - -

clockwise

23. The magnetic field is at right angles to the conductor, and there is a definite relation between the direction of the current flow and the direction of the associated magnetic field—if one is reversed, so is the other. If the direction of current flow, as shown in Figure 6-4, is reversed, the

direction of the magnetic field will be (clockwise/counterclockwise)

_____ (as viewed from the top).

- - - - - - - - - -

counterclockwise

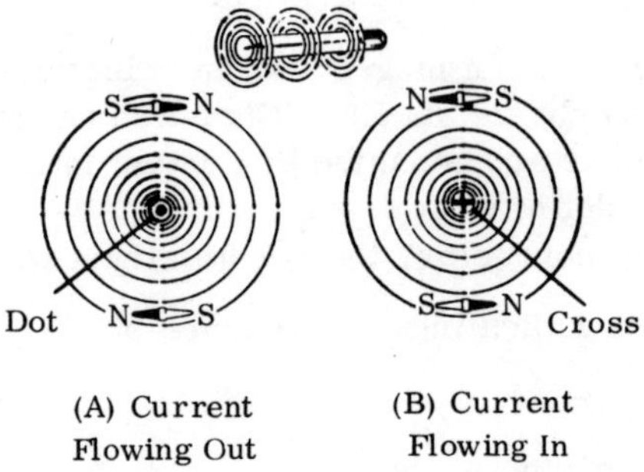

(A) Current
Flowing Out

(B) Current
Flowing In

Figure 6-5.   Cross-section of a
current-carrying conductor.

24. The relation between the magnetic field and the direction of current flow
is more easily visualized if the conductor is shown in cross-section with
the current thought of as flowing into or out of the paper. In Figure 6-5,
when the direction of current flow is out of the page (toward the reader),
this direction is indicated by a dot, representing the point of an arrow.
When the current flow is into the page (away from the reader), the cross,
representing the tail of an arrow, is used. Place your <u>left</u> hand over
View A so that your thumb is pointing up and your fingers are curved
around the cross-sectional view of the conductor. The curvature of your

fingers is (clockwise/counterclockwise) _____.

- - - - - - - - - -

clockwise

25. Now rest the tip of your <u>left</u> thumb on the cross in View B so that your
fingers are curved around the cross-sectional view of the conductor. The

curvature of your fingers is _____.

- - - - - - - - - -

counterclockwise

26. You have just demonstrated the <u>Left-Hand Rule for a Conductor</u>. This rule is used to determine the relation between the direction of the magnetic lines of force around a conductor and the direction of current flow along the conductor. If the conductor is grasped in the left hand with the thumb extended in the direction of electron flow (negative to positive), the fingers will be curved in the direction of the magnetic lines of force. (If the conductor is not insulated, do this <u>mentally</u>!) Figure 6-5 also illustrates the deflection of a compass needle in a magnetic field about a conductor. Which pole of the compass needle points in the direction of the

magnetic lines of force? _____.

- - - - - - - - - -

the north pole

27. How would you determine the direction of the magnetic lines of force about a current-carrying conductor if the direction of current is known?

_____

_____

- - - - - - - - - -

Grasp the conductor with the left hand so that the thumb points in the direction of current flow. The fingers will curve about the conductor in the direction of the magnetic lines of force.

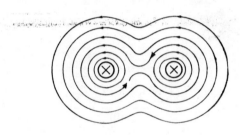

(A) Currents Flowing in
the Same Direction

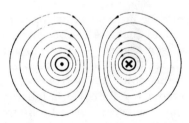

(B) Currents Flowing in
Opposite Directions

Figure 6-6. Magnetic field around two parallel conductors.

28. In View A of Figure 6-6, current in both conductors is flowing into the
page. When two parallel conductors carry current in the same direction,
the magnetic fields tend to encircle both conductors, drawing them together with a force of attraction. When two parallel conductors carry current in opposite directions, as in View B, the magnetic fields

(attract/repel) _____.

- - - - - - - - - -

repel

29. In View B, what is the physical effect of the magnetic fields on the conductors? _____

- - - - - - - - - -

The conductors tend to be pushed apart.

If you wish to take a break, do so now.

Magnetic Field of a Coil

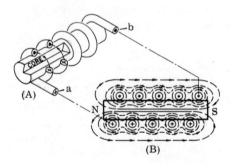

Figure 6-7. Magnetic field produced
by a current-carrying coil.

30. The magnetic field around a current-carrying wire exists at all points
along its length. The field consists of concentric circles in a plane perpendicular to the wire (see Figure 6-4). A partial cutaway view of a simple coil is shown in Figure 6-7, View A. View B is a complete cross-
sectional view of the same coil. Each cross-section of conductor (actually
a single conductor wound around a core) is labeled with a dot or a cross
to indicate the direction of current flow. In the top portion of the coil, the
current is flowing out of the paper. As indicated in Frame 28, when two
parallel conductors carry current in the same direction, the magnetic
fields tend to draw the conductors together. The same is true of any number of parallel conductors. In a coil, there is actually only one conductor,

but the turns are, in effect, parallel conductors. In the coil shown in Figure 6-7, the "parallel conductors" on the bottom half carry current in

a direction that is (the same as/opposite to) _____ that of the conductors in the top half.

- - - - - - - - - -

opposite to

31. There are two magnetic fields around the turns of wire that make up the

coil in Figure 6-7. The two fields (repel/attract) _____ each other.

- - - - - - - - - -

repel

32. When current is passed through the coiled conductor in Figure 6-7 the magnetic field of each turn of wire links with the fields of adjacent turns. The combined influence of all the turns produces a two-pole field similar to that of a simple bar magnet. As in all magnets, it has a north pole and a south pole. The north pole of the magnet produced by the current flow in the coil is, as in all magnets, the pole at which the flux lines

(enter/leave) _____ the magnet.

- - - - - - - - - -

leave

33. In Figure 6-5, it was shown that the direction of the magnetic field around a straight conductor depends on the direction of current flow through that conductor. Thus, a reversal of current flow through a conductor causes a reversal in the direction of the magnetic field that is produced. Reversal of current flow through a coil also causes a reversal of its two-pole field. In Figure 6-7, if the current flow through the coil is reversed, the

left end of the coil will become a (north/south) _____ pole.

- - - - - - - - - -

south

34. There is also a <u>left-hand rule for coils</u>, which is illustrated in Figure 6-8 on the next page. If you grasp a coil in the left hand, with the fingers "wrapped around" the coil in the direction of current flow, the thumb will point toward one of the poles.

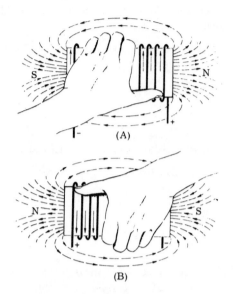

Figure 6-8. The left-hand rule
for coil polarity.

The thumb is pointing toward the (north/south) _____ pole.

- - - - - - - - - -

north

35. If a simple coil is grasped in the left hand so that the thumb points toward the north pole, the fingers will curve in a direction that is

(the same as/opposite to) _____ the direction of current flow.

- - - - - - - - - -

the same as

36. The left-hand rule for coils is used to determine the _____ of a coil.

- - - - - - - - - -

polarity

37. The strength, or intensity, of the magnetic field of a coil depends on a number of factors.

• As the <u>number of turns</u> of a conductor about a coil is increased, the field strength also increases.

- As <u>current flow</u> through the coil increases, field strength increases.

- Field strength depends on the <u>type of material in the core</u>. The better the core material can conduct magnetic lines of force (soft iron is an excellent conductor), the greater the field strength. (The measure of magnetic conduction is called "permeability," which will be discussed later.)

Coil A and Coil B are the same size and have the same type of core material. Each has the same type of wire wound about it, and each has the same amount of current flowing through the wire. Coil A, however, has 20 turns of wire, while Coil B has 30. Which coil has the greater

magnetic field strength? _____

- - - - - - - - - -

coil B

38. A relay is a type of electromagnet that is used to open or close one or more connections in a circuit. A certain relay's coil does not have sufficient field strength to operate its contacts, and the only thing that can be changed in the circuit is the source voltage. To make the relay work,

would you increase or decrease the source voltage? _____

- - - - - - - - - -

Increasing the source voltage would increase the current through the coil, thereby increasing the field strength to make the relay work.

39. If a core of soft iron is inserted in an air-core coil, will the magnetic

field become stronger or weaker? _____ Why? _____

_____

- - - - - - - - - -

Stronger; soft iron is a better conductor of magnetic lines of force than air.

There are equations for the precise determination of magnetic fields, but it is not necessary to study them now. We have seen that the core material affects the strength of a magnetic field. Now let us examine some of the properties of magnetic materials.

Properties of Magnetic Materials

40. When an annealed sheet steel core is used in an electromagnet it produces a stronger magnet than if a cast iron core is used. This is true because annealed sheet steel is more readily acted upon by the magnetizing force

of the coil than is hard cast iron. In other words, soft sheet steel is said
to have greater permeability because of the greater ease with which magnetic lines are established in it. Permeability is the relative ease with
which a substance conducts magnetic lines of force. The permeability of
air is arbitrarily set at 1. The permeability of other substances is the
ratio of their ability to conduct magnetic lines compared to that of air;
the figure is often much greater than 1. Here are the relative permeability figures for three magnetic materials: wrought iron, 1,500; sheet
steel, 2,310; cast iron, 600. (These figures are not constant but depend
on other factors beyond the scope of this book.) Which of the three materials has the greatest ability to conduct magnetic lines of force?

_____

- - - - - - - - - -

sheet steel

41. The permeability of magnetic materials varies with the degree of magnetization and with the type of material. A highly magnetized material (one
in which the flux density is great) cannot readily conduct magnetic lines
of force because it is already near saturation. That is, it has a great
concentration of lines of force already. Since its ability to conduct lines
of force is reduced, it has a lower permeability than the same material
when it is not so highly magnetized. Permeability of a magnetic material
that has not been magnetized is a certain figure that depends on the type
of material. When it begins to be magnetized, its permeability decreases
as its concentration of lines of force, or flux density, increases. There-

fore, permeability depends on what two factors? _____

_____

- - - - - - - - - -

the type of material and the flux density (degree of magnetization)

42. Hysteresis is another property of magnetic materials that should be considered when the core material of a coil is selected. Any magnetic material becomes magnetized to some degree when magnetic lines of force
pass through it. In the case of a coil, the flow of current through the coil
produces a magnetic field which results in the whole coil's becoming magnetic with a north pole and a south pole. When the current is reversed,
the polarity of the coil is also reversed.
     When current flows through the coil, a magnetic polarity is established.
If the coil has a core of some magnetic material, such as iron, the core
material will become magnetized and will show evidence of being magnetized even when the current is shut off.   For reasons that are beyond the
scope of this book, there is a lag in the reversal of magnetic polarity when
the current is reversed. This property that causes the magnetization to
lag is called hysteresis. The lag is caused by molecular friction, and

where there is friction of any kind, energy is dissipated. Energy dissipated through molecular friction is called <u>hysteresis loss</u>. Hysteresis loss, then, is a dissipation of energy that results from the tendency of magnetization to lag behind the force that produces it. The more easily a material can be magnetized, the shorter the lag will be. Would hysteresis loss be greater in a material that is hard to magnetize, or in one

that is easy to magnetize? _____

- - - - - - - - - -

Hysteresis loss is greater in a material that is hard to magnetize.

43. If the magnetization is reversed slowly, the energy loss may be negligible. But in the case of alternating current, the direction of current flow changes rapidly, as you will see later in the book. Therefore, hysteresis losses could be significant. It happens that some substances retain their magnetization more easily than others. Hard steel, for example, retains its magnetism better than cast iron. If current through a coil with a hard steel core is reversed, the core does not readily reverse its polarity (compared to cast iron). Its lag, in other words, is greater. Is hysteresis loss greater in hard steel or in cast iron? _____

- - - - - - - - - -

hard steel

44. The properties of permeability and hysteresis are both important in electricity, because they must be considered in designing an electrical circuit. It is important to keep energy losses to a minimum, and hysteresis loss is a type of energy loss. The type of magnet of most interest to us in the study of electricity is the electromagnet, which we will study in the next section. When possible, the core material selected for the coil of an

electromagnet should be a substance that has (high/low) _____ hysteresis.

- - - - - - - - - -

low

Electromagnets

45. An electromagnet is composed of a coil of wire wound around a core that is normally soft iron, because of its high permeability and low hysteresis. When direct current flows through the coil, the core will become magnetized with the same polarity that the coil would have without the core. If the current is reversed, what happens to the polarity of both the coil and

the core? _____

- - - - - - - - - -

Both are reversed.

46. When is an electromagnet a magnet? The other types of magnet we have mentioned — the natural magnet and the permanent magnet — are always magnets, unless something happens to destroy their magnetism. But the electromagnet is a temporary magnet. It is of great importance in electricity simply because the magnetism can be "turned on" or "turned off" at will. An example of this is the starter solenoid in your automobile. A solenoid is a current-carrying coil designed for a specific purpose. In your car, it is part of a relay that connects the battery to the induction coil, which generates the very high voltage needed to start the engine. The starter solenoid isolates this high voltage from the ignition switch. When no current flows in the coil, it is "air-core," but when the coil is energized, a movable soft-iron core does two things. First, the magnetic flux is increased because the soft-iron core is more permeable than the air core. Second, the flux is more highly concentrated. All this concentration of magnetic lines of force in the soft-iron core results in a very good magnet when current flows in the coil. But soft-iron loses its magnetism quickly when the current is shut off. The effect of the soft iron is, of course, the same whether it is movable, as in some solenoids, or permanently installed in the coil. An electromagnet, then, consists basically of a coil and a core. When is an electromagnet a magnet? _____

- - - - - - - - - -

When current flows in the coil.

47. The ability to control the action of magnetic force makes an electromagnet very useful in many circuit applications. You have probably noticed that either pole of an ordinary bar magnet will attract a magnetic material such as soft iron. The same is true of an electromagnet. The component that is to be attracted by the electromagnet is called the armature. One application of an electromagnet is illustrated in Figure 6-9, which shows a magnetic circuit breaker. Since a circuit breaker is a protective device that is designed to trip when current reaches its maximum permissible value, the adjustment of the armature is important.

Magnetic lines of force emanate from a magnet indefinitely, but the field becomes weaker as the distance from the magnet is increased. You might have noticed that a weak magnet must almost touch bits of metal before they are picked up, while a strong magnet operates over a greater distance. When an electromagnet is used as a circuit breaker, its armature (the part that unlatches the contacts) must be carefully adjusted so that it is attracted by the magnetic field, thus opening the breaker contacts, only when there is a need to protect the circuit. The magnetic field strength of an electromagnet increases as the current through its coil increases. Therefore, the armature can be adjusted so that its distance from the magnet is just great enough for it to be attracted only when the rated current of the circuit is exceeded. If it is set too close to the magnet, the circuit breaker might be tripped under normal operating conditions; while too great a distance would result in a failure to trip if the current load became too high.   Let us assume that a circuit breaker is designed to be tripped when the current in the circuit exceeds 20 a.   It actually trips, however, when the circuit current is only 18 a. Should the

armature be set closer to or farther from the magnet? _____

- - - - - - - - - -

It should be set farther from the magnet.   (It should not "feel" the magnetic attraction until the circuit current exceeds 20 a.)

48. The pair of contacts are normally latched closed by the armature, and current flows in the circuit. When the armature is pulled against the core of the electromagnet, the current in the circuit is interrupted because

_____ .

- - - - - - - - - -

the contacts are forced open by the action of a spring, breaking the circuit   (Note: The circuit breaker is reset manually.)

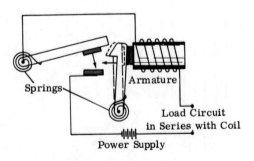

Figure 6-9. Magnetic circuit breaker.

Magnetic Shielding

49. In electricity, insulators are used to control voltage or restrict the flow of current. But there is no known insulator to control magnetic flux, the magnetic "equivalent" of current. Placing a nonmagnetic material in a magnetic field does little or nothing to change the flux. For example, a magnet can attract bits of iron quite well through glass, which is a good insulator for electric current. Stray magnetic fields, however, can influence the sensitive mechanism of an electric instrument and cause an error in its reading. The instrument can't be insulated against magnetic flux but it can be <u>shielded</u>. Shielding is the redirection of magnetic flux.

Why is it necessary to shield sensitive instruments? _____

_____

- - - - - - - - -

To protect them from magnetic influence, which can cause error. (If
you said it was because they can't be insulated against magnetism, you
are also correct.)

50. A material that can be easily magnetized, such as soft iron, is useful for
shielding. If a case of such a material (which is said to have <u>high perme-
ability</u>) is built around a sensitive instrument, most of the magnetic flux
in its vicinity will go into the case instead of into the instrument. A mag-

netic shield (blocks/redirects/destroys) _____ magnetic flux.

- - - - - - - - - -

redirects

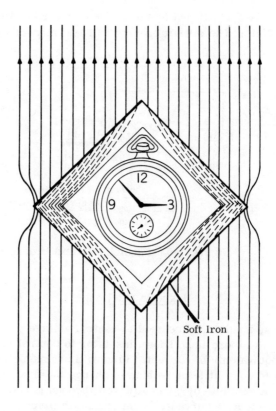

Figure 6-10. A magnetic shield.

51. Figure 6-10 on the preceding page illustrates how a magnetic shield
    works. How is the watch shielded from the magnetic flux? _____

    _____

    _____

- - - - - - - - - -

Something like: The watch is enclosed in a case of soft iron, which has
high permeability. (You might also have said that the case was easily
magnetized.) The magnetic flux is directed around the watch because
it is directed into the case instead.

    In this chapter you have learned that, as in electricity, magnets have
polarity and conform to the laws of attraction and repulsion. You have
become acquainted with natural and permanent magnets and have learned
the special properties of electromagnets.
    You have learned that a magnetic field is always associated with elec-
tric current, and you have become acquainted with the laws governing the
relationship of current to its magnetic field.
    The properties of magnetic materials that are most important in elec-
tricity—permeability and hysteresis—have been described, as well as the
special properties of electromagnets.
    Since sensitive instruments cannot be insulated against magnetic fields,
they must be shielded. You have learned how this is accomplished.
    When you feel that you have a good understanding of the material cov-
ered in this chapter, go on to the Self-Test.

Self-Test

The following questions will test your understanding of Chapter Six. Write
your answers on a separate sheet of paper and check them with the answers
provided following the test.

1. What are the points of maximum attraction on a magnet?

2. What is the law of attraction and repulsion as it is related to magnetism?

3. What is a magnetic field?

4. What are the three general groups of magnets?

5. What is the relationship between flux density and magnetic field strength?

6. How can you determine the direction of the magnetic field about a conductor when the direction of current flow is known?

7. When the left-hand rule is applied to a coil, the thumb points to which
   pole?

8. Name three ways the magnetic field strength of a current-carrying coil can be increased.

9. Permeability depends on what two factors?

10. What is hysteresis?

11. When does an electromagnet actually function as a magnet?

12. When an electromagnet is used as a circuit breaker, what factor determines the amount of current flow in the coil that will cause the armature to be attracted?

13. How is a sensitive instrument shielded from stray magnetic fields?

Answers

If your answers to the test questions do not agree with the ones given below, review the frames indicated in parentheses after each answer before you go on to the next chapter.

1. The (magnetic) poles.   (1)

2. Unlike poles attract; like poles repel.   (5)

3. The flux lines, or magnetic flux, along which a magnetic force acts.   (8)

4. Natural magnets, permanent magnets, and electromagnets.   (11)

5. The two are directly proportional. As flux density increases, field strength also increases.   (16)

6. Grasp the conductor in the left hand so that the thumb points in the direction of current flow. The fingers will curve around the conductor in the direction of the magnetic field.   (29)

7. North.   (37)

8. (a) Increase the number of turns of wire in the coil; (b) increase the current flow through the coil; and (c) select a core material with higher permeability.   (40)

9. The type of material and the flux density.   (44)

10. The lag between magnetization and the force that produces it.   (45)

11. When current flows in the coil.   (49)

12. The maximum permissible current that should flow in the protected circuit.   (50)

13. The instrument is enclosed in a highly permeable material that will conduct the magnetic lines of force around it.   (54)

CHAPTER SEVEN

Introduction to Alternating-Current Electricity

In direct-current electricity, you became familiar with electron flow in one direction. For example, in a battery, which is a common source of d-c electricity, electron flow is from the negative terminal through the circuit to the positive terminal.

In alternating current, the subject of this chapter, the flow of electrons is not continuous in one direction. The electrons are made to move first in one direction and then in the other. The direction of current flow reverses many times a second.

When you have finished this chapter you will be able to:

- explain how an alternator generates an a-c voltage;

- explain frequency and state the factors that affect it;

- analyze the sine waveform of a-c voltage or current;

- make the mathematical conversions between maximum and effective values of voltage or current; and,

- use vectors to add values of a-c voltage or current.

The Basic Alternating-Current Generator

1. The only practical way to produce alternating current is by means of a generator, which is a machine that converts mechanical energy into electrical energy. The electrical energy might be in the form of either direct current or alternating current, so the alternating-current generator is usually called an <u>alternator</u>, the term we shall normally use in this book. The basic components of an alternator are an armature, about which many turns of conductor are wound, which rotates in a magnetic field, and some means of delivering the resulting alternating current to an external circuit. We will go into the construction in more detail later, but first, let's look at the theory of operation.

During your study of magnetism in Chapter Six, you learned that a current-carrying conductor produces a magnetic field around itself. A magnetic field may also, under certain conditions, induce an electromotive force (emf) in a conductor. One such condition exists when there is

relative motion between the magnetic field and the conductor.   (If either
the conductor or the field moves, there is relative motion between the
two.) It is not very practical to move a magnetic field while the conductor remains stationary. How else might we achieve relative motion be-

tween the magnetic field and the conductor? _____

_____

- - - - - - - - - -

We could move the conductor within the magnetic field (or equivalent
words).

2. When a conductor is moved within a magnetic field, it cuts lines of force
    as it moves. When the conductor cuts lines of force, current flows, pro-
    vided there is a complete path for current flow in the conductor. This
    process, called electromagnetic induction, is one way to cause current

    flow in a conductor. What is electromagnetic induction? _____

_____

- - - - - - - - - -

The effect that causes current flow in a conductor moving across
magnetic lines of force (or similar wording).

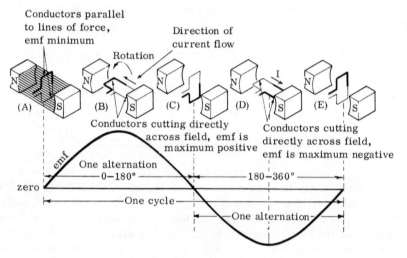

Figure 7-1. Basic alternator.

Refer to Figure 7-1 for Frames 3 through 20.

3. In Figure 7-1, a suspended loop of wire (conductor) is shown as it is ro-
    tated in a counterclockwise direction through the magnetic field that exists
    between the poles of a permanent magnet. For ease of explanation, the

loop has been divided into a dark and a light half. In this discussion, consider only the dark half. At position A, the conductor is parallel to the
lines of force. It will not begin to cut the lines of force until it moves a-
way from that position. The S-shaped line in the bottom half of Figure
7-1 represents varying values of voltage (emf) induced in the conductor
as it rotates through the magnetic field. (This line is called a sine wave,
which we will discuss later.) If the conductor is cutting no lines of force,
as in position A, what is the value of emf induced in the conductor? _____

- - - - - - - - - -

zero

4. Let's assume that the conductor is rotating at a constant speed. It cuts
more and more lines of force as it moves away from position A because
it is cutting more directly across the field. (That is, the angle of "cut"
is approaching the perpendicular.) As it moves toward position B, what
happens to the emf existing in the conductor at each instant of time—does

it increase or decrease? _____

- - - - - - - - - -

It increases.

5. As the loop continues its rotation toward position C, it cuts fewer and
fewer lines of force during each increment of time. Therefore, what

happens to the induced voltage? _____

- - - - - - - - - -

It decreases.

6. As the armature of an alternator moves through one complete rotation,

what must occur for maximum voltage to be generated? _____

_____

- - - - - - - - - -

The conductor(s) must cut a maximum number of lines of force.

7. That portion of the sine wave above the zero line represents positive voltages, while the portion below the zero line represents negative voltages.
Therefore, at some point between position A and position E the current
flowing in the conductor changes direction. Periodic changes of direction
are a characteristic of alternating current. Can you identify at least one

point at which current flow changes direction? _____

What kind of current characteristically has periodic changes of direction?

_____

- - - - - - - - - -

It changes direction at positions C, E, and A; alternating current

8. Remember that the sine wave represents constantly changing values of voltage. The height above the zero line of any point on the sine wave represents the relative value of voltage at that point in the rotation of the conductor, compared with voltages generated at other points. As the conductor cuts more lines of force, the induced voltage increases. Of the positions pictured in Figure 7-1, the conductor cuts the greatest number of lines of force at what positions? _____

- - - - - - - - - -

B and D

9. Position C is a point of (maximum/zero) _____ voltage.

- - - - - - - - - -

zero

10. A and E are also points of zero voltage. At what position has the conductor returned to its original starting point? _____

- - - - - - - - - -

E   (So positions A and E are equivalent.)

11. A sine wave of voltage represents all the <u>instantaneous</u> values of voltage induced in a conductor as it moves through one complete rotation. A rotation of the conductor through 360 degrees is one <u>cycle</u>. A sine wave of voltage, then, might also be called a cycle of voltage. The positive part of the cycle is between what two positions? _____

- - - - - - - - - -

A and C

12. In Figure 7-1, the induced voltage is zero when the conductor is parallel to the lines of force, as in position A. The instantaneous voltages vary as the conductor rotates, but they are all positive between positions A and C. The current flow changes direction when the conductor reaches position C, so all the instantaneous voltages between positions C and E are

(positive/negative) _____.

- - - - - - - - - -

negative

13. In your own words, describe a sine wave as it relates to the armature of

an alternator. _____

_____

- - - - - - - - - -

A sine wave is a waveform that represents all the instantaneous values
of current or voltage as the armature of an alternator moves through
one complete rotation.

14. The conductor moves through a complete circle of 360 degrees during one
    rotation. The portion of the sine wave representing rotation of the con-
    ductor between 0 and 180 degrees is designated one <u>alternation</u>. The por-
    tion of the sine wave between 180 and 360 degrees also represents one
    alternation. An alternation can be thought of as the portion of the cycle
    during which current flow is in one direction. The current flow changes

    direction how many times during one complete cycle? _____

- - - - - - - - - -

This was not intended to be a trick question. If you considered that the
current flow reversed at position A, then it changes direction twice
during the cycle. Otherwise, it changes direction only once, at position
C. The point is that current flows in one direction during the first half
of a cycle and then in the other direction for the second half.

15. The wave of induced voltage goes through one complete cycle between po-
    sitions A and E. If the loop is rotated at a steady rate, and if the strength
    of the magnetic field is uniform, the number of cycles per second and the
    effective voltage will remain at fixed values. Continuous rotation will
    produce a series of sine-wave voltage cycles; in other words, an alternat-
    ing-current (a-c) voltage. In this way mechanical energy is converted into

    what other kind of energy? _____

- - - - - - - - - -

electrical energy

Frequency

16. The rotating loop in Figure 7-1 is actually wound on an armature. For
    simplicity, a single loop is shown, but the armature may have any number
    of loops or coils wound on it. The armature is moved at a constant speed,
    because we want the same number of sine waves, or cycles, of voltage to

be generated each second.   The number of cycles per second is the frequency of the alternator.

   If the armature is made to rotate faster, will the frequency increase

or decrease?   _____

- - - - - - - - - -

It will increase.

17. You have probably heard of the term "60-cycle voltage" in reference to
     the ordinary alternating-current electricity used in homes and industry
     in the United States.   This term refers to the frequency of the current or
     voltage; that is, the number of sine waves that are generated each second.
     Several years ago, by international agreement, the old term "cycles per
     second," abbreviated "cps," was discontinued in favor of "hertz," abbre-
     viated "Hz," in honor of Gustav Hertz, a German physicist.   We will use
     the old term, "cycles per second," only when it seems necessary to make
     a point clear.

        In the United States, the standard frequency is 60 cycles per second,
     or 60 Hz, while in many other parts of the world, it is 50 Hz.   (A home
     tape cassette recorded in Europe and played in the U.S. might sound a
     bit odd for that reason, but that is another story.)   Frequency (f) is relat-
     ed to time.

        If a single-loop armature makes 60 complete rotations in one second,

     what is its frequency?   _____

- - - - - - - - - -

     60 Hz

18. If the conductor is rotated faster in the magnetic field of an alternator,

     what will happen to the frequency of the voltage?   _____

- - - - - - - - - -

     It will increase.

19. The frequency (number of complete cycles per second) is the same as
     the number of rotations per second if the magnetic field is produced by
     only two poles.

        On the next page, however, Figure 7-2 illustrates a four-pole alterna-
     tor; and in this alternator, one revolution of the armature produces two
     complete cycles of voltage.

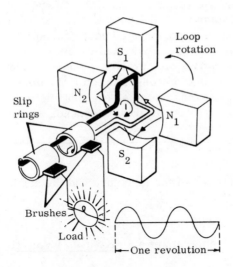

Figure 7-2.   A four-pole basic alternator

If a four-pole alternator has an armature rotation of 30 revolutions per second, what is the frequency of the resulting output voltage? _____

- - - - - - - - - -

60 Hz

20. An alternator produces how many cycles of voltage for each pair of poles producing its magnetic field? _____

- - - - - - - - - -

one

21. Figure 7-2 shows some other components of an alternator: brushes and slip rings. It would be pretty awkward to have a lot of wiring twisting around itself as the armature rotates, so a system of slip rings and brushes provides the means of transferring the electric current from the generator to the outside circuitry. The slip rings are attached to the armature, while the brushes (usually made of carbon) are connected to the outside circuits and held against the slip rings by spring tension. There is contact, but not mechanical connection. If a spring were to break, allowing a brush to move away from the slip ring, the result would be a(n) (open/closed) _____ circuit.

- - - - - - - - - -

open

22. The <u>period</u> of an a-c voltage is the time for one complete cycle, or 1/f. The period of ordinary 60-Hz voltage, for example, is 1/60 of a second, since 60 sine waves are generated each second. It is not necessary to calculate the periods of a-c voltages in this book, but the frequency is essential in calculating other values, such as inductive and capacitive reactance, which will be discussed in later chapters. The output voltage of an alternator depends on the strength of its magnetic field. The output <u>frequency</u> of an operating alternator depends on what two factors?

- - - - - - - - - -

The speed of rotation of the armature and the number of poles creating the magnetic field.

Analysis of a Sine Wave of Voltage

23. In earlier chapters, it was a simple matter to add various quantities of resistance, voltage, current, and power, because in direct current, all currents and voltages are <u>in phase</u>. That is, current and voltage are either present or not present, depending on whether a circuit is open or closed. In alternating current, on the other hand, current and voltage are often <u>out of phase</u>. Other circuit values (capacitance and inductance, which will be discussed later) might cause current to reach its peak ahead of or behind voltage. Phase will be discussed in detail later; for now, it is enough to know that a-c voltages and currents sometimes cannot be added by simple arithmetic. To solve problems involving sine-wave voltages and currents, you must understand <u>vectors</u>.

    A vector is a line representing both <u>magnitude</u> and <u>direction</u>. If we say the wind is blowing at a speed of 10 mph, we are making a statement about magnitude only. A wind out of the northwest at 10 mph, however, is described in terms of both magnitude and direction. If we drew a line representing only wind speed, it would <u>not</u> be a vector, because it would not represent direction. (Similarly, a line representing direction only is not a vector.) If we drew a line representing wind speed <u>and</u> direction, the line would be a vector. A vector of any value represents what two characteristics? _____

- - - - - - - - - -

magnitude; direction

24. In the next drawing, the north-south and east-west axes show direction, while each mark crossing an axis indicates an increment of 5 mph. The arrow is a vector representing wind velocity.

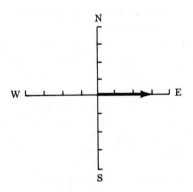

What are the magnitude and direction of the wind in this drawing?

- - - - - - - - -

The wind is out of the west at 15 mph. (Wind is generally thought of as <u>from</u> a direction, but this need not concern you here. If you stated the correct direction and magnitude, you answered correctly.)

25. Draw an arrow to indicate a wind out of the northeast at 10 mph. Each increment represents 5 mph. (You need not be exact; just approximate.)

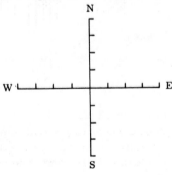

- - - - - - - - -

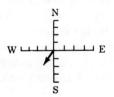

26. Direction does not have to mean north, south, etc. It can also mean negative or positive, or perhaps the number of degrees away from a given point in a circle. In the drawing below, the horizontal line represents degrees of rotation of an armature. The armature of an alternator rotates in a circle, which of course has 360 degrees. As the armature rotates, the conductors cut more and more lines of force at each instant of time, and the induced voltage rises from zero to some maximum positive value, then decreases to zero, then rises to a negative peak value, and finally decreases to zero as the armature reaches its starting point. The vertical lines represent instantaneous values of induced voltage, with positive values shown above the line and negative values shown below the line. Maximum voltage is induced at one-quarter of the full circuit of rotation (90 degrees) and three-quarters of the full circle (270 degrees). Zero voltage is induced at the starting point (0 or 360 degrees) and halfway through the circle (180 degrees). If you connect the ends of all the vertical lines with a single line (smoothed into a curve, since not all values are shown),

what is the result? _____

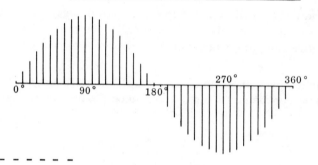

- - - - - - - - -

a sine wave

27. Some vectors, such as E_M in Figure 7-3 on the following page, may be rotated like the spokes of a wheel to generate angles. (The vectors representing instantaneous voltages in the sine wave are <u>not</u> rotating vectors.) Positive rotation is counterclockwise and generates positive angles. Negative rotation is clockwise and generates negative angles. The vertical projection of a rotating vector may be used to represent the voltage induced in an armature at any given instant. In Figure 7-3, the line E_M is a rotating vector, and the dashed line e is the vertical projection representing the instantaneous voltage when the angle θ (theta) reaches 60 degrees. The cross-section of the armature (conductor) is shown, with its positions numbered 0/12 (the starting and ending points are the same) through 11. Angle θ is 60 degrees when the armature is at position 2.

What is angle θ at position 3? _____

- - - - - - - - - -

90 degrees

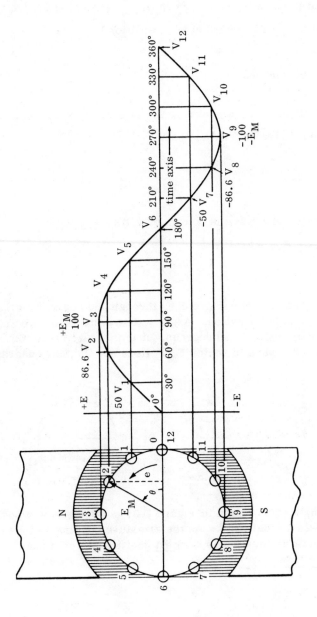

Figure 7-3.   Generation of sine-wave voltage.

Refer to Figure 7-3 for Frames 27 through 41.

28. The vector E_M represents the maximum voltage that can be induced in the armature for a given magnetic field strength. E_M represents e, the instantaneous voltage, only when the conductor is cutting the maximum number of lines of force in a given time. The only positive value of e that is

equal to E_M is induced at what position? _____

- - - - - - - - - -

3, or 90 degrees

29. The values of e are shown on the sine wave as V_1, V_2, etc. The maximum

positive value of e is labeled V____ .

- - - - - - - - - -

V_3

30. V_3 on the sine wave is how many volts? _____

- - - - - - - - - -

100 v.

31. The rotating vector E_M moves through 360 degrees as the armature completes one full rotation. Angle θ, however, is always an acute angle—its maximum value is 90 degrees. It is measured from the zero (base) line of the sine wave to vector E_M. Here are the four quadrants of the circle.

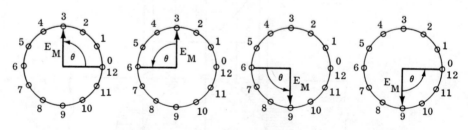

Angle θ, as shown in Figure 7-3, is 60 degrees at position 2 and 90 degrees at position 3. As the armature continues to rotate through the magnetic field, what is the next position at which angle θ is 60 degrees?

- - - - - - - - - -

position 4

32. Write in the values of angle   for the following positions of the armature.

    6 ____°   8 ____°   10 ____°

    7 ____°   9 ____°   12 ____°

- - - - - - - - -

6, 0°; 7, 30°; 8, 60°; 9, 90°; 10, 60°; 12, 0°

33. The value of e is maximum when angle θ is ____° and minimum (zero) when angle θ is ____°.

- - - - - - - - -

90°; 0°

34. Angle θ must be 90 degrees or less, but the armature rotates through 360 degrees. Angle θ is 30 degrees when the armature is at ____, ____, ____, and ____.

- - - - - - - - -

30, 150, 210, 330

35. The voltage wave form produced as the armature moves through 360 degrees of rotation is called a sine wave because the instantaneous voltage e is related to the sine (abbreviated sin) of the angle θ. (Sine is a trigonometric function that expresses the ratio between the length of the side opposite a given angle in a right triangle to the hypotenuse of that right triangle.) The sine curve is a graph of the equation

$$e = E_M \sin \theta,$$

where e is the instantaneous voltage, E_M is the maximum voltage, and θ is the angle of the generator armature. You have already learned that $e = E_M$ when $\theta = 90$ degrees. Therefore, you don't need trigonometric tables to discover the sine of a 90-degree angle. You merely have to ask, "What value must be substituted for sin θ to preserve the equation $e = E_M$?" The sine of a 90-degree angle must be ____.

- - - - - - - - -

1

36. The instantaneous voltage e cannot be greater than E_M, and angle θ is never greater than 90 degrees. Therefore, the sine of any other value of angle θ must be less than ____.

- - - - - - - - -

1

37. Any value e may be found by multiplying E_M by sin θ. (The sines of angles between 0 and 90 degrees may be found in a table of trigonometric functions.) The sine of 60 degrees is 0.8660. If E_M = 100 v., what is the induced voltage when the armature of a generator is at 120 degrees (angle θ = 60)? _____

- - - - - - - - - -

86.6 v.

38. When the armature in Figure 7-3 is at any point between 0 and 180 degrees, e is positive. The value of e is negative when the armature is at any point between _____ and _____ degrees.

- - - - - - - - - -

180 and 360 degrees

39. When the armature is at 60 or 120 degrees and E_M = 100 v., the value of e is (+86.6/−86.6) _____ v.

- - - - - - - - - - -

+86.6

40. When the armature is at 240 or 300 degrees and E_M = 100 v., the value of e is _____ v. (Include the polarity.)

- - - - - - - - - -

−86.6

41. Sin 30 = 0.5000 and E_M = 100 v. What is the value of e when the armature is at 150 degrees? _____ When the armature is at 330 degrees?

_____

- - - - - - - - - -

+50.0; −50.0

We have seen that the magnitude of the voltage generated by an alternator changes from instant to instant as the armature rotates in the magnetic field. By connecting all the instantaneous vectors, we can see that a symmetrical sine wave results from one complete rotation, or cycle. The information gained so far is helpful in understanding the nature of alternating current, but we need some method of translating the instantaneous voltages into a constant value that is useful in solving problems. We will discuss such a method in the next section.

This is a convenient place to take a break.

Effective Value of Alternating Current or Voltage

42. The instantaneous value of voltage constantly changes as the armature of an alternator moves through a complete rotation. Since current varies directly with voltage, according to Ohm's Law, the instantaneous changes in current also result in a sine wave whose positive and negative peaks and intermediate values can be plotted exactly as we plotted the voltage sine wave. However, instantaneous values are not useful in solving most a-c problems, so an <u>effective</u> value is used. The effective value of an a-c voltage or current of sine waveform is defined in terms of an <u>equivalent heating effect</u> of a direct current. Heating effect is independent of the direction of current flow. Since all instantaneous values of induced voltage are somewhere between zero and E_M (maximum, or peak voltage), the effective value of a sine wave voltage or current must be greater than

zero and less than what? _____

- - - - - - - - - -

E_M (the maximum, or peak voltage)

43. An alternating current of sine waveform having a maximum value of 14.14 amperes produces the same amount of heat in a circuit having a resistance of one ohm as a direct current of 10 amperes. Since this is true, we can work out a constant value for converting any peak value to a corresponding effective value. This constant is represented by X in the simple equation below. Solve for X to three decimal places.

$$14.14X = 10$$

$$X = \_\_\_\_\_$$

- - - - - - - - - -

0.707

44. The insulation of a conductor must be designed to withstand the <u>peak</u> voltage, not merely the effective voltage. A voltmeter, however, reads the effective value. Let us assume that a certain type of insulation is intended to withstand 1,000 v. How can we find the highest effective voltage (which is the value we can read on a voltmeter) that should occur in the circuit?

_____

- - - - - - - - - -

Multiply 1,000 by 0.707.

45. The effective value is also called the root-mean-square (rms) value because it is the square root of the average of the squared values between zero and maximum. (You may be relieved to know that you don't have to remember the math.) The effective value of an a-c current is stated in terms of an equivalent d-c circuit. What phenomenon is used as the standard of comparison? _____

- - - - - - - - - -

The heating effect of the current.

46. A-c voltmeters and ammeters (which read voltage and current, respectively) are designed to read the effective value. If the peak a-c voltage across a resistor is 100 v., what is the rms voltage? _____

- - - - - - - - - -

70.7 v.   (100 x 0.707)

47. The peak current in an a-c circuit is 84.8 ma. An ammeter would indicate (to the nearest milliampere) _____ ma.

- - - - - - - - - -

60   (84.8 x 0.707)

48. As we mentioned earlier, the insulation for conductors must be designed for the maximum voltage. A little arithmetic will show you that the peak voltage is about 1.41 times the effective voltage. When a circuit is designed, if the effective voltage between two conductors in a cable is 440 v., what maximum voltage must the insulation be able to withstand? _____

- - - - - - - - - -

About 620 v.   (440 x 1.41)

49. Effective voltage (or current) in an a-c circuit is _____ times the peak value.

- - - - - - - - - -

0.707

50. Peak voltage (or current) in an a-c circuit is _____ times the effective value.

- - - - - - - - - -

1.41

Combining A-C Voltages

Figure 7-4. Combining a-c voltages. Refer to Figure 7-4 for Frames 51 through 67. Two armatures, several waveforms, and a vectorial diagram are presented in this illustration. You will not understand the entire illustration immediately, but it is necessary to show several relationships, which will become clear as you work through the frames.

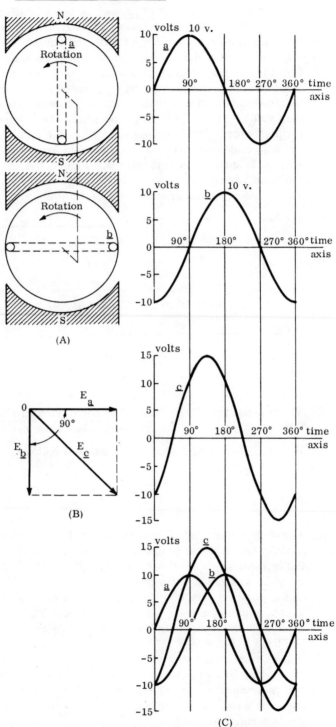

51. It is sometimes necessary to calculate the combined effect of a-c voltage or currents that are <u>out of phase</u>, as you will discover when you study inductance and capacitance later in this book. That is, the peaks of various voltages or currents occur at different times. For example, one voltage might be at maximum while another is at zero or some other value. Vectors may be used to combine such voltages or currents. Figure 7-4 shows two armatures (labeled <u>a</u> and <u>b</u>) mounted on a common shaft in an alternator and connected in series. The two armatures are shown separately to help you understand what is happening. The armatures are rotating counterclockwise. The voltages induced in the two armatures are out of phase, because maximum voltage is induced in armature <u>a</u> at the point where zero voltage is induced in armature <u>b</u>. Figure 7-4 shows the two armatures at the point where that condition exists. Look at the positive peaks of the sine waveforms for armatures <u>a</u> and <u>b</u> (the top two waveforms). The peaks are how many degrees apart? _____

- - - - - - - - - -

90 degrees

52. The voltages induced in the two armatures are how many degrees out of phase? _____

- - - - - - - - - -

90 degrees

53. The maximum voltage induced in each armature is 10 v. The sine wave for armature <u>a</u> starts at zero degrees of its own rotation. If armature <u>a</u> is at the zero point, what is its induced voltage? _____ v.

- - - - - - - - - -

zero

54. Remember that armature <u>b</u> is mounted on the same shaft as armature <u>a</u>, but the two armatures are physically placed 90 degrees apart. We are using armature <u>a</u> as a reference in Figure 7-4. The top waveform, sine wave a, has reached its peak positive value of 10 v. when armature <u>a</u> is in the position shown. The time axis of each waveform is laid out in degrees, using armature <u>a</u> as a reference. The position of armature <u>b</u> is always 90 degrees behind armature <u>a</u>, so when it is at zero degrees on the time axis, it is at 270 degrees of its own rotation. When armature <u>a</u> is at zero degrees <u>of its own rotation</u>, it is also at zero degrees on the time axis, because it is used as the reference. Remember that armature <u>b</u> lags armature a by 90 degrees. When armature <u>a</u> is at 180 degrees on the time axis, armature <u>b</u> is, too, but it occupies some other position <u>in its own rotation cycle</u>.

When armature a is at 180 degrees, armature b has rotated through how many degrees of its own cycle? _____

- - - - - - - - - -

90

55. When armature a is at zero degrees, its induced voltage is also zero. At that same point along the time axis, maximum (positive/negative) _____ voltage is induced in armature b.

- - - - - - - - - -

negative

56. Armature a (leads/lags) _____ armature b by 90 degrees.

- - - - - - - - - -

leads

57. The third waveform in Figure 7-4 shows the <u>combined</u> induced voltages of armatures a and b. It represents the actual output of the alternator whose two armatures are shown in view A. What is the output of the alternator at the zero point of the time axis? (Indicate polarity.) _____

- - - - - - - - - -

−10 v.   (Note: This is easy to see, because armature a is at zero—no induced voltage—while armature b is at 270 degrees, which is −10 v.)

58. Compare waveforms a, b, and c in Figure 7-4. These waveforms show, respectively, the voltage induced in armature a, the voltage induced in armature b at the same time, and the combined voltage output of armatures a and b. Look at the waveforms at 90 degrees on the time axis and answer the following questions.
    (a) When armature a is at 90 degrees, armature b is at how many

        degrees of its own rotation? _____
    (b) What is the voltage induced in armature a at 90 degrees on the

        time axis? (Indicate polarity.) _____
    (c) What is the voltage induced in armature b at 90 degrees on the

        time axis? _____
    (d) What is the output of the alternator at 90 degrees on the time

        axis? (Indicate polarity.) _____

- - - - - - - - - -

(a) zero; (b) +10 v.; (c) zero; (d) +10 v.

59. Each of the two armatures has a peak voltage of 10 v. What is the <u>effec-</u>
<u>tive</u> voltage induced in each armature? _____

- - - - - - - - - -

7. 07 v.   (0. 707 x 10)

60. The two effective voltages cannot be combined by arithmetic because they
are 90 degrees out of phase. However, they can be added by the use of
vectors. View B of Figure 7-4 shows a vectorial representation of the
voltages induced in armatures <u>a</u> and <u>b</u>. E_a represents the effective voltage induced in armature <u>b</u>. <u>The vectors represent the effective equiva-</u>
<u>lents of the first two waveforms. Do not try to relate them to the third</u>
<u>waveform.</u> Remember that the vectors are generated <u>counterclockwise.</u>

E_a (leads/lags) _____ E_b by _____ degrees.

- - - - - - - - - -

leads; 90 degrees

61. The combined effect of the two voltages represented by vectors E_a and E_b
is shown by vector E_c. If the coordinates of the vectors (E_a and E_b) in
View B were marked off in units to represent voltages, you could just
measure E_c (using the same units) to obtain the effective output of the alternator, but this would be laborious. Such vectors are normally solved
by the mathematics of right triangles. First you have to identify the right
triangle that is the basis of your mathematics. In View B of Figure 7-4,
vectors E_a and E_b are the coordinates that represent specific voltages:
the outputs of armature <u>a</u> and armature <u>b</u>. Vector E_c is the resultant, or
the vector that represents the combined output of the two armatures.
Since vectors E_a and E_b are both generated from the same point, one of
them (it doesn't matter which) must be moved to form a triangle with the
other two vectors. The dashed lines represent the positions of vectors
E_a and E_b after they are moved. The dashed line parallel to each vector
is the "new" position of that vector.

  The Pythagorean theorem states that the square of the hypotenuse of
any right triangle is equal to the sum of the squares of the other two sides.
This is stated mathematically as

$$a^2 + b^2 = c^2.$$

Take the square root of each side of the equation:

$$\sqrt{a^2 + b^2} = \sqrt{c^2}.$$

If two sides of a right triangle are 3 and 4 units, the length of the hypotenuse can be found by substituting:

$$\sqrt{9 + 16} = \sqrt{c^2}$$
$$5 = c$$

The length of vector E_c can be found by the use of the Pythagorean theorem, since it is known that E_a and E_b are each 7.07. Solve for E_c. (Note: The 7.07 is actually rounded off a little, so round off each number in your solution to the nearest whole number.) _____

- - - - - - - - -

E_c = 10 v. $(\sqrt{50 + 50} = 10)$

62. If the sides of a right triangle forming the right angle are equal, as E_a and E_b are equal, there is an easier solution. The length of one of the equal sides multiplied by 1.414 (the square root of 2) is equal to the hypotenuse. (We won't explore the reason in this book.) Using this solution, if E_a and E_b are each 12 v., E_c is _____ v. (Round off your answer.)

- - - - - - - - -

17 v.   (16.968)

63. View C in Figure 7-4 shows all three waveforms (a, b, and a + b) on the same time axis. The instantaneous output of the alternator is about +15 v. when both a and b are at about +7.5 v. This point is approximately how many degrees on the time axis? _____

- - - - - - - - -

$135°$

64. At what points on the time axis is the instantaneous output of the alternator zero? _____

- - - - - - - - -

approximately 45 and 225 degrees

In this chapter you have become acquainted with the basic theory of alternating current. You have learned how the sine-wave voltage output of an alternator is generated and how output voltages are combined vectorially. You have examined the relationship between peak voltage and effective voltage and have had some practice in solving problems involving these values.

This chapter provided some of the basic information that you will need in later chapters, when you study some of the special characteristics of a-c circuits.

When you feel that you understand the material in this chapter, turn to the Self-Test.

Self-Test

The following questions will test your understanding of Chapter Seven. Write your answers on a separate sheet of paper and check them with the answers provided following the test.

1. What happens when a conductor (which provides a complete path for current flow) is moved in a magnetic field?

2. What is electromagnetic induction?

3. As the armature of an alternator moves through one complete rotation, what must occur for maximum voltage to be generated?

4. What characteristic of alternating current distinguishes it from direct current?

5. What is a sine wave?

6. An alternator converts _____ energy into _____ energy.

7. What is frequency?

8. What is the modern unit of frequency? What is its abbreviation?

9. What are the two factors that affect the frequency of the voltage generated by an alternator?

10. What two values are represented by a vector?

11. What directions are indicated by the vectors that describe a sine wave?

12. As an armature moves through one complete cycle of rotation, a maximum positive voltage of 50 v. is induced at 90 degrees of rotation away from zero. As rotation continues:
    (a) At how many degrees away from zero will the voltage induced in the armature next be zero?
    (b) At how many degrees away from zero will the induced voltage be −50 v. ?

13. The maximum voltage generated by an alternator is 200 v. What is the effective voltage?

14. The rms voltage generated by an alternator is 500 v. What is the peak voltage?

15. Two armatures, x and y, are mounted on the same shaft of an a-c generator. When armature x is at 180 degrees, armature y is at 145 degrees. Armature x (leads/lags) _____ armature y by _____ degrees.

16. How can the combined value of out-of-phase voltages or currents be calculated?

17. The following vectors represent the effective values of two out-of-phase voltages generated by counterclockwise rotation of the armatures.

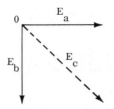

Vector (E_a/E_b) _____ leads by _____ degrees.

18. Vectors E_a and E_b in question 17 each represent 4. 244 v. The resultant vector E_c represents _____ v.

Answers

If your answers to the test questions do not agree with the ones given below, review the frames indicated in parentheses after each answer before you go on to the next chapter.

1. A voltage is induced in the conductor.   (1)

2. The effect that causes current flow in a conductor moving across magnetic lines of force.   (2)

3. The conductor(s) must cut a maximum number of lines of force.   (3-7)

4. It periodically changes direction.   (7)

5. A waveform that represents all the instantaneous values of current or voltage as the armature of an alternator moves through one complete rotation.   (8-12)

6. mechanical; electrical   (15)

7. The number of sine waves generated each second.   (17)

8. The hertz; Hz   (17)

9. The speed of rotation of the armature and the number of poles producing the magnetic field.   (22)

10. Magnitude and direction.   (23)

11. Positive and negative.   (26)

12. (a) 180 degrees; (b) 270 degrees   (27-30)

13. 141. 4 v.   (44)

14. 707 v.   (48)

15. leads; 35   (51-57)

16. By the use of vectors.   (51)

17. E_a; 90   (60)

18. 6   (61, 62)

CHAPTER EIGHT
Inductance

So far the circuits we have studied have been <u>resistive</u>. That is, resistors presented the only opposition to current flow. Two other phenomena—inductance and capacitance—exist in direct-current circuits to some extent, but they are of major importance in alternating-current circuits. Both inductance and capacitance present a kind of opposition to current flow that is called "reactance," which you will study in Chapter Ten. Before we examine reactance, however, we must first study inductance (the subject of this chapter) and capacitance (the subject of Chapter Nine).

Inductance is the property of an electric circuit that opposes any <u>change in the current</u> through that circuit. That is, if the current increases, a self-induced voltage opposes this change and delays the increase. If the current decreases, a self-induced voltage tends to aid (or prolong) the current flow, delaying the decrease. Thus, current can neither increase nor decrease as fast in an inductive circuit as it can in a purely resistive circuit.

This effect becomes very important in a-c circuits, because it affects the <u>phase</u> relationships between voltage and current. You learned in Chapter Seven that voltages (or currents) can be out of phase if they are induced in separate armatures of an alternator. In that case, the voltage and current <u>generated by each armature</u> were in phase. When inductance is a factor in a circuit, the voltage and current generated by the <u>same</u> armature are out of phase. We shall examine these phase relationships later in this book. Your objective in this chapter is to understand the nature and effects of inductance in an electric circuit.

The opposition to the change of current is essentially an effect of electromagnetic induction, or induced electromotive force (emf). In Chapter Seven, you learned that voltage is induced in a conductor when it is moved through a magnetic field. The same thing happens when a magnetic field is moving across a conductor. It is this relative motion between the field and the conductor that produces a self-induced voltage in a conductor. A magnetic field builds up when current begins to flow and collapses when it is shut off. In either case, the field <u>moves</u>, and this effect is the subject of this chapter.

When you have finished this chapter you will be able to:

- explain the factors that affect inductance;

- describe the growth and decay of current in a resistive-inductive circuit;

- distinguish between self-inductance and mutual inductance;

- describe the effects of inductance in an electric circuit;

- calculate the induced voltage when the inductance is known;
- calculate the L/R time constant of a circuit; and,
- calculate the total inductance of a circuit.

Self-Inductance

1. As you have already learned, current flow in a conductor always produces a magnetic field surrounding, or linking with, the conductor. Inductance is a factor in a circuit even when a conductor is perfectly straight, because the magnetic field builds up, or expands, when current begins to flow and deteriorates, or collapses, when current is shut off. Inductance is much more significant when the conductor is in the form of a coil, because each turn of the coil, with its associated magnetic field, affects all other nearby turns, as we shall see later. When the current changes (this change may be an increase, a decrease, or a change of direction), the magnetic field also changes, and an electromotive force (emf) is induced in the conductor. This emf is called a self-induced emf because it is induced in the same conductor carrying the current. The self-induced emf is caused by the magnetic field as it moves across a conductor. Self-

   inductance results from what characteristic of electric current? _____

   ___ ___ ___ ___ ___ ___ ___ ___ ___ ___

   Any current has an associated magnetic field.

2. When current flows in a conductor, what must be the relationship between the magnetic field and the conductor to cause a self-induced emf?

   ___ ___ ___ ___ ___ ___ ___ ___ ___ ___

   The magnetic field must cut, or move across, the conductor.

3. The strength of a magnetic field about a conductor is directly proportional to the amount of current flow in the conductor. When current flow in-

   creases, what happens to the magnetic field strength? _____

   ___ ___ ___ ___ ___ ___ ___ ___ ___ ___

   It increases.

4. The self-induced emf in a conductor is directly proportional to the strength of the magnetic field that causes it. As the current flow in a conductor in-

   creases, what happens to the magnetic field strength? _____

What happens to the self-induced emf? _____

- - - - - - - - - -

It increases; it also increases.

5. When current flows in a conductor, how is a self-induced emf produced
   in the conductor? _____

- - - - - - - - - -

The associated magnetic field moves across (cuts) the conductor,
inducing an emf.

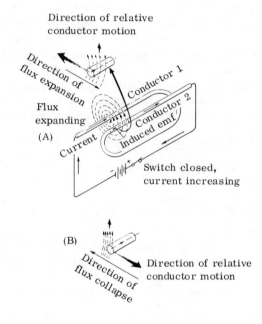

Figure 8-1. Self-inductance.

Refer to Figure 8-1 for Frames 6 through 12.

6. When current flows in a conductor, it produces a magnetic field (flux) that
   cuts across the conductor and induces an emf. The induced emf causes
   current flow that is in the <u>opposite direction</u> to the current flow already
   present in the conductor. This effect is summarized by Lenz's Law:
   <u>The induced emf in any circuit is always in a direction to oppose the effect</u>
   <u>that produced it</u>. Don't try to visualize a <u>flow</u> of emf; rather, think of the
   <u>current flow</u> resulting from the emf. Inductance is increased by shaping
   a conductor so that the electromagnetic field around each portion of the
   conductor cuts across some other portion of the same conductor. This is
   shown in its simplest form in Figure 8-1. A length of conductor is looped

so that two portions of the conductor lie adjacent and parallel to one another. These portions are labeled Conductor 1 and Conductor 2. When the switch in the circuit is closed, electron flow through the conductor establishes a typical concentric field around <u>all</u> portions of the conductor. For simplicity, the field is shown in a single plane that is perpendicular to both conductors. Although the field originates simultaneously in both conductors, it is considered here as originating in Conductor 1, and its effect on Conductor 2 will be shown. When the switch in the circuit is closed,

the flux begins to (expand/contract) _____.

- - - - - - - - -

expand

7. When the flux expands, there is relative motion between it and the conductor, and the effect is the same as if the conductor were moving. You have already learned that when a conductor moves in a magnetic field, a voltage (emf) is induced in the conductor. A small section of Conductor 2 is shown in View A of Figure 8-1. The direction of the expanding flux from <u>Conductor 1</u> is shown as well as the direction of the relative motion of Conductor 2. The current flow in Conductor 2 that is produced by the battery is shown in the larger drawing. (You can locate the arrow that indicates this current flow by following the arrows from the negative terminal of the battery.) A dashed-line arrow in the small segment of Conductor 2 shows the current flow resulting from the induced emf. The induced emf causes current flow in Conductor 2 that is in a direction that is (the same as/

opposite to) _____ the current flow produced by the battery.

- - - - - - - - -

opposite to

8. The effect of inductance in a circuit is to <u>oppose any change in the current</u> through that circuit. Current in a purely resistive circuit is normally thought of as starting and stopping instantly when the circuit is closed or opened. To understand the effect of inductance, you must realize that changes in current flow are not quite instantaneous. To visualize this, review what happens when an inductance is present in the circuit, as it is in Figure 8-1. When the switch is closed, current begins to flow in the circuit. The current flows in all parts of the circuit, of course, but examine the portions of the circuit labeled Conductor 1 and Conductor 2. When current begins to flow in Conductor 1, the resulting magnetic field begins to expand, cutting across Conductor 2. An emf is induced in Conductor 2, and this self-induced emf causes an <u>additional</u> current flow that is in a direction opposite to the main current flow. Thus, when the switch

is closed, the inductance of Conductor 1 acts to (aid/oppose) _____ the buildup of current in Conductor 2.

- - - - - - - - - -

oppose

9. Although the effect is not illustrated in Figure 8-1, a magnetic field also begins to expand around Conductor 2 when the circuit switch is closed. As current from the battery begins to flow in Conductor 1, the expanding magnetic field from Conductor 2 induces an emf. How does this emf affect the

current flow in Conductor 1? _____

_____

- - - - - - - - - -

It opposes the current flow in Conductor 1.

10. View B shows what happens when the circuit switch is opened. When the current that produces the magnetic field is turned off, the field begins to collapse. Therefore, its direction of movement is reversed, and the apparent direction of movement of Conductor 2 in that field is also reversed. Now the direction of current flow in Conductor 2 is in the same direction as that of the battery current. Although the battery current tries to stop (because the switch is opened), the collapsing magnetic field induces an emf in Conductor 2 that tries to keep the current moving. This continues only for the tiny length of time during which the magnetic field is collapsing, but the effect is real. Therefore, how does inductance act to influence

the current in the circuit? _____

- - - - - - - - - -

It opposes any change in current flow.

11. How does inductance act on the <u>battery</u> current, (a) when the switch is closed and, (b) when the switch is opened? _____

_____

- - - - - - - - - -

(a) Inductance opposes current flow; (b) inductance aids current flow.

12. When the circuit switch is closed, there is a <u>change</u> in current from zero to its normal value. When the switch is opened, there is again a change back to zero. It is important to note that the voltage of self-inductance opposes <u>both</u> changes in current. It delays the initial buildup of current by opposing the battery voltage, and it delays the breakdown of current by exerting an induced voltage that causes current flow in the same direction

as the battery current. When current flows in any circuit, how does self-inductance affect the current flow? _____

_____

- - - - - - - - - -

Self-inductance always opposes any change in current in the circuit.

If you intend to take a break pretty soon, this is a good place to stop.

13. The unit for measuring inductance, L, is the henry (named for an American physicist, Joseph Henry), abbreviated h. You have now encountered five parameters, or things measured, in electric circuits. (You will learn others later.) Complete the table below.

Parameter	Mathematical Symbol	Unit	Abbreviation
voltage	E	volt	v.
current	___	___	___
resistance	___	___	___
power	___	___	___
inductance	___	___	___

- - - - - - - - - -

current, I, ampere, a.; resistance, R, ohm, Ω; power, P, watt, w.; inductance, L, henry, h.

14. The henry is a large unit of inductance, so you need to learn two other abbreviations used with inductance. "Milli-," as you have learned, means a thousandth; that is, 1/1000, or 0.001. One milliampere equals 0.001 a. A common unit of inductance is the millihenry, abbreviated mh. One millihenry equals _____ h.

- - - - - - - - - -

0.001; 1/1000

15. An easier way to express 0.001 h. is _____.

- - - - - - - - - -

1 mh.

16. An even smaller unit of inductance is also common. You have already learned that "micro-" means a millionth, or 0.000001. The Greek letter μ (pronounced mu) is used to represent "micro-," so a millionth of a henry is called a _____, abbreviated _____.

- - - - - - - - - -

microhenry; μh

17. Inductance opposes a <u>change</u> in current. The Greek letter Δ (pronounced delta) means "a change in." The expression for "a change in time" is Δt.

What do you think would be the expression for "a change in current"? ___

- - - - - - - - - -

ΔI

18. An inductance of 1 henry exists if an emf of 1 volt is induced when the current is changing at the rate of 1 ampere per second. The relationships of induced voltage, inductance, and rate of change of current with respect to time can be stated mathematically:

$$E = L \frac{\Delta I}{\Delta t}$$

where E is the induced emf in volts, L is the inductance in henries, and ΔI is the change in current in amperes that occurs in a given amount of time (Δt). You can find the value of the voltage induced in a conductor such as that shown in Figure 8-1 if you know the other values. If all units are of the same order of magnitude, such as milli- or micro-, the induced voltage will be of the same order of magnitude; that is, it will have the same prefix. For example, to solve for induced voltage where the inductance is 5 μh, the change in current is 4 microamperes, and the time required to effect the change is 2 microseconds, substitute those values in the equation:

$$E = 5 \text{ x } (4 \div 2)$$
$$= 5 \text{ x } 2$$
$$= 10$$

The induced voltage is 10 (volts/microvolts/millivolts) _____.

- - - - - - - - - -

microvolts

19. Try working another problem yourself. The inductance is 10 mh, the change in current is 2 ma., and the change in time is 1 millisecond. What is the induced voltage? _____

- - - - - - - - - -

20 mv.   (20 millivolts)   (E = 10 x (2 ÷ 1) = 20)

20. L = 2 µh;   I = 3 microamperes; and Δt = 2 microseconds.   E = _____

- - - - - - - - - -

3 microvolts

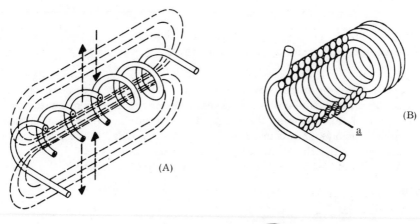

(A)

(B)

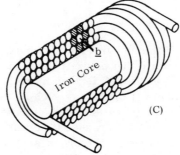

(C)

Figure 8-2.   Coils of various inductors.

Refer to Figure 8-2 for Frames 21 through 26.

21. Many things affect the self–inductance of a circuit. One important factor
is the degree of <u>linkage</u>, or interaction, between the circuit conductors
and the associated magnetic flux. In a straight length of conductor, there
is very little flux linkage between one part of the conductor and another.
Therefore, its inductance is extremely small. Conductors become much
more inductive when they are wound into coils, as shown in Figure 8-2.
This is true because there is maximum flux linkage between the conductor
turns, which lie side by side in the coil. Of the three coils (A, B, and C)
shown in Figure 8-2, which one do you think has the poorest flux linkage?

_____

- - - - - - - - - -

The one in View A.   (The next frame explains why.)

22. The coil in View A is a poor inductor compared to the others, because its turns are widely spaced, thus decreasing the flux linkage between its turns. (Remember that a magnetic field becomes weaker as it moves out in space.) How could you increase the flux linkage of the coil in View A?

_____

- - - - - - - - - -

By decreasing the spacing between its turns.

23. A more inductive coil is shown in View B.   The turns are more closely spaced, and since there are two layers, the turns are linked with a greater number of flux loops as the magnetic field moves out, as indicated by the dashed arrows shown in View A.   In other words, there are more opportunities for the flux to induce an emf in the conductor.
     The turn labeled a in View B is directly adjacent to how many other

turns? _____

- - - - - - - - - -

four

24. Most of the turns of the coil in View B are adjacent to four other turns, so there is great interaction between the turns and much less flux is "wasted." The coil in View C is even more inductive because it is wound in three layers. Some of the turns, such as the one labeled b, are directly adjacent to six other turns (shaded). Increasing the number of layers in which the coil is wound increases its cross-sectional area, which improves lateral flux linkage. The cross-sectional area of a coil is an important factor in its inductance. To make a coil more inductive would you

increase or decrease its cross-sectional area? _____

- - - - - - - - - -

Increase it.

25. We have seen that the inductance of a coil is affected by the number of turns, the spacing between the turns, and the cross-sectional area of the coil. The spacing of the turns and the cross-sectional area are actually considered together as one factor: the ratio of the cross-sectional area to the length of the coil. Summarize the two major factors affecting in-

ductance that you have learned so far. _____

_____

- - - - - - - - - -

(1) the number of turns; (2) the ratio of the cross-sectional area to the length of the coil.

26. One other important factor affecting inductance is illustrated by the coil in View C of Figure 8-2, which is the most inductive of the three shown. Not only does it have the greatest number of turns and the greatest ratio of cross-sectional area to length, but it has an iron core. (The other two coils have air cores.) As you learned in Chapter Six, permeability (symbol μ) is related to the core material of a coil. An iron core is highly permeable, which greatly increases the inductance of the coil in View C. As the permeability of the core material is increased, what happens to

    the inductance? _____

    - - - - - - - - - -

    It increases.

27. The Greek letter mu (μ) stands for two things in electricity. (1) As the symbol for a characteristic of a magnetic material, it represents

    _____. (2) As a numerical prefix, it is an

    abbreviation for the prefix _____, which means _____.

    - - - - - - - - - -

    (1) permeability; (2) micro-; one millionth (or 0.000001)

28. There are several equations in advanced electricity that take permeability into account. While you will not study them in this book, you should know that the permeability of a magnetic material (μ) is assigned a numerical value that represents its relative ability to conduct magnetic lines of force as compared with air, which is arbitrarily assigned a μ of 1. The magnetic materials commonly used in electrical applications have μ values that are much greater than 1. In an air-core coil, what is the value of

    μ? _____

    - - - - - - - - - -

    1

29. The inductance of a coil increases very rapidly as the number of turns is increased. It also increases as the coil is made shorter, the cross-sectional area is made larger, or the permeability of the core is increased. The four major factors affecting inductance are:

    (1) _____

    (2) _____

(3) _____

(4) _____

– – – – – – – – –

(1) the number of turns;   (2) the length of the coil;   (3) the cross-sectional
area of the coil;   (4) the permeability of the core

(Note: If you listed the ratio of the cross-sectional area to the length,
you accounted for two of the factors.)

30. As you have learned, any electrical device has resistance, but sometimes
resistors are used in a circuit for the simple purpose of limiting current
flow. There is the same relationship between the terms <u>inductance</u> and
<u>inductor</u>. Any current-carrying conductor has inductance, but a device
manufactured to introduce inductance into a circuit is called an inductor.
For example, there are inductors in a radio circuit that help to tune the
radio to a certain station. As you might expect, there is a schematic
symbol for inductance (whether it is planned or unplanned). Here is a
simple schematic diagram that includes some symbols you have already
learned as well as the symbol for an inductor.

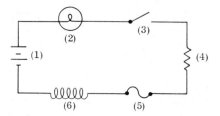

The symbols in the above schematic are numbered. Write the meaning
of each symbol.

(1) _____   (4) _____

(2) _____   (5) _____

(3) _____   (6) _____

– – – – – – – – – –

(1) battery;   (2) lamp;   (3) switch;   (4) resistor;   (5) fuse;   (6) inductor

Growth and Decay of Current in an R–L Series Circuit

31. If a battery is connected across an inductor, the current builds up to its
final value at a rate that is determined by the battery voltage and the circuit resistance. The current buildup is gradual because of the opposing
emf (counter emf) resulting from the self-inductance of the coil.

How does the counter emf affect the buildup of current? _____

- - - - - - - - - -

It opposes the buildup of current.

32. When the current starts to flow, the magnetic lines of force move out, cut the turns of wire on the inductor, and build up a counter emf. What do

you think this counter emf opposes? _____

- - - - - - - - - -

The emf of the battery.   (If you said "battery current," you are also correct.)

33. This opposition causes a delay in the time it takes the current to build up to a steady value. Later, when the battery is disconnected, the lines of force collapse, again cutting the turns of the inductor and building up an

emf. What does this emf tend to do? _____

- - - - - - - - - -

prolong the current flow

34. Although the analogy is not exact, electrical inductance is somewhat like mechanical inertia. A large truck begins to move forward in a very low gear, because the inertia of a heavy weight at rest must be overcome. As the heavy truck begins to pick up speed, the driver changes to successively higher gears as the load becomes easier to move. The reverse occurs when the driver wishes to stop the truck. He must gear down to overcome the tendency of forward inertia to keep the load moving forward. In the case of inductance, it is electrical "inertia" that must be overcome. Figure 8-3, on the next page, shows a circuit that includes two switches, a battery, and a voltage divider containing containing a resistor (R) and an inductor (L). (The voltage dividers we have studied so far have consisted of resistors only; however, an inductor may be used as a component.) Switches S_1 and S_2 are "ganged," as indicated by the dashed line, so that when one is closed, the other is opened at exactly the same instant. Such an arrangement is called an R-L (resistive-inductive) series circuit. The source voltage of the battery is applied across the R-L combination

(the resistor and inductor) when (S_1/S_2) _____ is closed.

- - - - - - - - - -

S_1

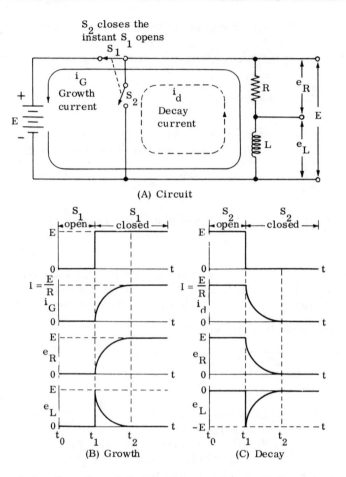

(A) Circuit

(B) Growth   (C) Decay

Figure 8-3.   Growth and decay of current in an R-L series circuit.

Refer to Figure 8-3 for Frames 34 through 43.

35. As soon as S_1 is closed, as shown in View A of Figure 8-3, a voltage (E) appears across the R-L combination. A current attempts to flow, but the inductor opposes this current by building up a counter emf. At the first instant S_1 is closed, the counter emf (sometimes called "back emf") exactly equals the battery emf, and its polarity is opposite. Under this condition, will current flow in resistor R? _____

- - - - - - - - - -

no

36. Because no current can flow when the counter emf is exactly equal to the battery voltage, no voltage is dropped across R. View B shows this situation. The growth current (i_G), voltage across R (e_R), and voltage

across L (e_L) are all plotted along a time line. S_1 is first closed at time ($t_0/t_1/t_2$) _____ .

- - - - - - - - -

t_1

37. At time t_1, the growth current i_G (the current through R) is (zero/maximum) _____ , e_R is (zero/maximum) _____ , and e_L is (zero/maximum) _____ .

- - - - - - - - - -

zero; zero; maximum

38. At t_1, all of the battery voltage is dropped across L. Since the sum of all voltage drops in a series circuit must be equal to the source voltage, there is no voltage dropped across R. As current overcomes the opposition of the inductance and starts to flow, a voltage (e_R) appears across R; and e_L, the voltage across L, is reduced by the same amount. As i_G increases, (e_L/e_R) _____ also increases.

- - - - - - - - - -

e_R

39. The growth current i_G reaches maximum shortly before time _____ .

- - - - - - - - - -

t_2

40. The curves in View B show that e_L finally reaches zero when i_G reaches maximum. At this point, all of the source voltage is dropped across the _____ .

- - - - - - - - -

resistor

41. Under the steady-state condition (when there is no counter emf and "normal" current flows in the circuit), only the resistor limits the size of the current. Any conductor is considered to have some resistance, but wire resistance is usually ignored. Therefore, the resistance of the coil in Figure 8-3 is regarded as zero. The inductance of the coil opposed the growth of current in View A of Figure 8-3. View C shows that it also opposes the decay of current when the battery switch S_1 is opened. S_1 is

opened, and S_2 is closed, at time $(t_0/t_1/t_2)$ _____.

- - - - - - - - - -

t_1

42. When S_2 is closed, the battery voltage is removed from the circuit, and another circuit is provided for the flow of the decay current i_d. When the battery voltage is removed from the circuit, the magnetic field of L collapses, and lines of force again cut the turns of the coil. The induced

voltage e_L is maximum at time (t_1/t_2) _____, as shown in View C. (Note the new position of the zero line, which was moved for easier comparison of the growth and decay curves.)

- - - - - - - - - -

t_1

43. At the first instant S_2 is closed, e_L is essentially the same value as the battery voltage, which is now disconnected. The decay current i_d flows through R in the same direction as did current i_G when the battery was in the circuit. At time t_1, e_R is maximum, but it rapidly falls to zero as

e_L reaches zero shortly before time _____.

- - - - - - - - - -

t_2

44. The growth of current in a circuit when the circuit is completed by a switch closure and the decay of current when a switch is opened are almost instantaneous. We have traced the growth and decay of current to show how an inductance opposes <u>any</u> change in current. When a circuit is closed, inductance tries to keep current from flowing. Its opposition is shortlived, however, because a <u>moving</u> magnetic field is necessary for a counter emf to be produced. Once the circuit current reaches its steady state, the field no longer moves, and no counter emf is produced.

When a circuit is opened, inductance again opposes a change in current flow and tries to keep it flowing by means of the counter emf, which is now produced by the collapsing magnetic field. But the opposition to a decrease in circuit current is only momentary, because there is no motion of the magnetic field (in fact, it does not exist) once it completely collapses.

How does inductance affect circuit current when a switch is first

closed to complete a circuit? _____

_____

- - - - - - - - - -

Inductance opposes the buildup of current in the circuit.

45. How does inductance affect circuit current when a switch is opened?

- - - - - - - - -

Inductance tries to keep the current flowing.

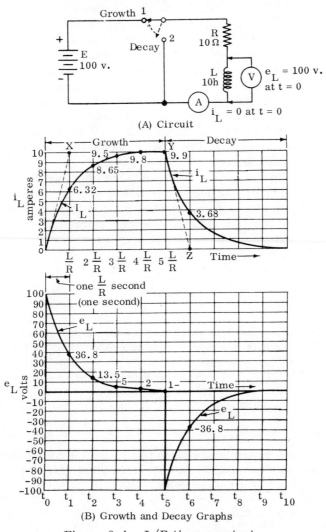

(A) Circuit

(B) Growth and Decay Graphs

Figure 8-4.   L/R time constant.

Refer to Figure 8-4 for Frames 46 through 54.

L/R Time Constant

46. The time required for the growth or decay of current in an R-L circuit is important in many electronic circuits. These applications are beyond the scope of this book, but if you plan to go on to advanced electricity, you

will want to learn the fundamentals of the L/R time constant.

As the curves in Figure 8-3 indicate, growth and decay are rapid at first, and then the curves flatten out. The top graph of Figure 8-4 shows the growth of current through the inductor (i_L), while the bottom graph shows the corresponding decay of induced voltage (e_L). After several intervals of time—which are labeled L/R, 2L/R, etc. on the i_L graph and t_1, t_2, etc. on the e_L graphs—the rate of growth or decay has slowed down to the point of insignificance. The first interval is the one from which the expression "time constant" is derived. We normally think of a constant as an unchanging value in an equation. In this case, the constant has to be calculated; it is only the significant percentage of maximum current or voltage that is constant.

The time required for the current through an inductor to increase to 63.2 percent of its maximum value or to decrease to 36.8 percent is known as the L/R time constant of the circuit. You can see, by comparing the two graphs in Figure 8-4, that when i_L is 63.2 percent of maximum, e_L is 36.8 of maximum.

Compare the two graphs in terms of the intervals of time. At time t_1 on the e_L graph, e_L has decreased to 36.8 percent as i_L has increased to 63.2 percent of maximum, or 6.32 a. The increment of time during which this 63.2 percent of maximum was achieved is found by dividing

_____ by _____.

- - - - - - - - - -

L by R (or inductance by resistance)

47. If L is in henries and R is in ohms, t (time) is in seconds. If L is in microhenries and R is in ohms, t is in microseconds. If L is in millihenries and R is in ohms, t is in _____.

- - - - - - - - - -

milliseconds

48. R in the L/R equation is always in ohms, and the time constant is of the same order of magnitude as L. Here are three useful relations used in calculating the L/R time constant:

$$\frac{L \text{ (in henries)}}{R \text{ (in ohms)}} = t \text{ (in seconds)}$$

$$\frac{L \text{ (in millihenries)}}{R \text{ (in ohms)}} = t \text{ (in milliseconds)}$$

$$\frac{L \text{ (in microhenries)}}{R \text{ (in ohms)}} = t \text{ (in microseconds)}$$

In Figure 8-4, $R = 10\Omega$ and $L = 10$ h. How long does it take i_L to reach 63.2 percent of its maximum value? _____

- - - - - - - - - -

1 second

49. If $L = 15$ mh and $R = 5\Omega$, how long does it take e_L to decay to 36.8 percent of its maximum value? _____

- - - - - - - - - -

3 milliseconds

50. Given the values of L and R, how would you find the time between t_0 and t_1? _____

- - - - - - - - - -

By dividing L by R.

51. The current through the inductor (i_L) in Figure 8-4 takes 1 microsecond to grow from 0 to 6.32 a. How long does it take i_L to grow from 0 to 9.8 a. ? _____

- - - - - - - - - -

4 microseconds   (At the end of 4 L/R intervals on the graph, the current i_L is 9.8 a.)

52. During the time i_L increases to 6.32 a., as shown in Figure 8-4, e_L decreases to _____ v.

- - - - - - - - - -

36.8   (This is the value at t_1 on the bottom graph, which corresponds to L/R on the top graph.)

53. The time constant of an L-R circuit is always expressed as a ratio between _____ and _____.

- - - - - - - - - -

inductance (or L) and resistance (or R)

54. What is the L/R time constant of an R-L circuit? _____

_____

- - - - - - - - - -

The time required for the current through an inductor to increase to 63.2 percent or to decrease to 36.8 percent of its maximum value.

Mutual Inductance

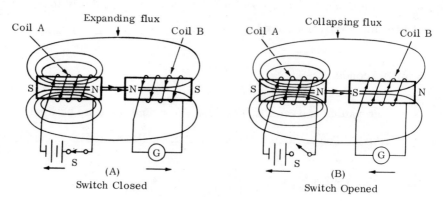

Figure 8-5. Mutual inductance.

Refer to Figure 8-5 for Frames 55 through 63.

55. Only one coil or conductor is involved in self-inductance. Whenever <u>two</u> coils are located so that the flux from one coil links with the turns of the other, a change of flux in one coil will cause an emf to be induced in the other. The two coils have <u>mutual inductance</u>. The amount of mutual inductance depends on the relative position of the two coils. If the coils are separated by a considerable distance, the amount of flux common to both coils is small, and the mutual inductance is low. On the other hand, if the coils are close together, so that nearly all the flux of one coil links the turns of the other, the mutual inductance is high. If we wind the two coils on the same iron core, would the mutual inductance be increased or decreased? _____

- - - - - - - - - -

increased

56. The two coils shown in Figure 8-5 are placed close together with their north-south axes in the same plane (that is, so that a single imaginary line could be extended through both coils). Coil A is connected to a battery through switch S, and Coil B is connected to a galvanometer (one kind of meter that measures current). When the switch is closed (View A), there is a momentary movement of the galvanometer needle to indicate current flow. What does this indicate? _____

_____

- - - - - - - - - -

That there is an induced current in Coil B.

57. The magnetic field associated with a conductor expands (moves out from its source) when current is turned on, then collapses (moves in toward its source) when current is turned off. In other words, the field is _moving_ as it expands or collapses. The current is induced in Coil B because the magnetic field generated by Coil A is (expanding/collapsing)

_____.

- - - - - - - - - -

expanding

58. When the current in Coil A reaches a steady value, the associated magnetic field is no longer expanding. Will the galvanometer now indicate current flow? _____

- - - - - - - - - -

no

59. When does current flow in Coil B? _____

_____

- - - - - - - - - -

Any time the magnetic field of Coil A is moving (either expanding or collapsing).

60. In View B, the switch is opened. The galvanometer will momentarily indicate current flow in Coil B. Why? _____

_____

- - - - - - - - - -

because the magnetic field of Coil A is collapsing

61. When current begins to flow in Coil B, an expanding magnetic field cuts the turns of Coil A. This causes a current opposite to battery current to be induced in Coil A. Thus, the buildup of current in Coil A is somewhat

_____.

- - - - - - - - - -

delayed (or opposed)

62. When the switch is opened in the circuit of Coil A, current does not stop flowing instantly because the collapsing field of Coil B tends to _____ the current flow in Coil A.

- - - - - - - - - -

prolong (or aid)

63. There would still be mutual inductance between the two coils in Figure 8-5 if they were placed side by side instead of end to end; but the degree of inductance would be different, because the magnetic fields would cut the turns of the coils in a different manner. For simplicity, we have discussed current flow in Coil B as it is related to the magnetic field of Coil A. Naturally, Coil A also has self-inductance because of its own current flow.

Remember that the magnetic field associated with Coil A induces a voltage (and thus causes current flow) in Coil B only when the magnetic field of Coil A is expanding or collapsing. The power source for Coil A is a battery, so the current must be turned on or off before there is mutual inductance. If the power source were a-c, the current would be changing direction constantly, and the field would be alternately expanding and collapsing as long as current flowed. As long as the magnetic field has motion in either direction, there is mutual inductance. Explain, in terms of the magnetic field of Coil A, how a voltage is induced in Coil B.

_____

_____

- - - - - - - - - -

A voltage is induced in Coil B any time there is motion of the magnetic field associated with Coil A, whether it is expanding or collapsing.

Calculation of Total Inductance

64. If inductors in series are well shielded, or located far enough apart to make the effects of mutual inductance negligible, the total inductance is calculated in the same manner as for resistances in series; you merely add them:

$$L_t = L_1 + L_2 + L_3 \ldots \text{(etc.)}$$

What is the total inductance in a series circuit containing three coils (totally shielded) whose values are 50 μh, 30 μh, and 20 μh? _____

- - - - - - - - - -

100 μh.

65. If there is no mutual inductance between coils in a parallel circuit, the total inductance is again calculated in the same manner as for resistances in parallel:

$$\frac{1}{L_t} = \frac{1}{L_1} + \frac{1}{L_2} + \frac{1}{L_3} \ldots \text{(etc.)}$$

A portion of a circuit contains three totally shielded inductors in parallel. The values of the three inductances are: 5 mh, 10 mh, and 30 mh.

What is the total inductance? _____

- - - - - - - - -

3 mh   Solution: $\dfrac{1}{L_t} = \dfrac{1}{5} + \dfrac{1}{10} + \dfrac{1}{30}$

$= 0.2 + 0.1 + 0.033$

$= 0.333$ (approximately)

$L_t = \dfrac{1}{0.333}$

$= 3$ mh

66. Two inductances in parallel (with no mutual inductance) have values of 3 μh and 6 μh. What is L_t? (Hint: Use the product-over-sum method.)

_____

- - - - - - - - -

2 μh   (18 ÷ 9)   (Note: If you study advanced electricity, you will need to know the effect of mutual inductance in solving for total inductance in both series and parallel circuits, but we will not attempt to calculate that here.)

In this chapter, you have learned how inductance opposes any change in current flow, through the principles of self-inductance, and how, through mutual inductance, current can be made to flow in a circuit that has no power source of its own.

You have learned how the <u>motion</u> of a magnetic field associated with an inductor causes a voltage to be induced in either the same inductor or some other inductor within the field.

You have learned to calculate the induced voltage when the inductance is known, the L/R time constant of a circuit, and the total inductance of both series and parallel inductors.

When you feel that you understand the material in this chapter, go on to the Self-Test.

Self-Test

The following questions will test your understanding of Chapter Eight. Write your answers on a separate sheet of paper and check them with the answers provided following the test.

1. Self-inductance results from what characteristic of electric current?

2. When current flows in a conductor, how is a self-induced emf produced in the conductor?

3. How does inductance act to influence the current flowing in a conductor?

4. For inductance, what are (a) the mathematical symbol, (b) the unit of measurement, and (c) the abbreviation of the unit of measurement?

5. Assume that an inductance of 5 μh exists when the current in a conductor is changing at the rate of 12 microamperes every 3 microseconds. What voltage is induced in the conductor?

6. List the four major factors affecting the inductance of a coil and state whether an increase in each factor will increase or decrease inductance.

7. Draw the schematic symbol for an inductor.

8. How does inductance affect circuit current in each of the following cases? (Include the action of the magnetic field in your answers.)
   (a) When a switch is first closed to complete a circuit.
   (b) When a switch is first opened to interrupt current flow.

9. The inductance of an R-L circuit is 15 mh and the resistance is 3Ω. What is the L/R time constant of the circuit?

10. The current through an inductor in an R-L circuit reaches 40 percent of its maximum value in 1 microsecond, 63.2 percent in 2 microseconds, and 95 percent in 6 microseconds. What is the L/R time constant of the circuit?

11. In the following diagram, there will be mutual inductance between Coil A and Coil B for a short time when the switch is closed and again when the switch is opened.

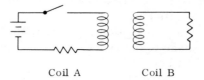

Coil A   Coil B

Explain, in terms of the magnetic field of Coil A, how a voltage is induced in Coil B.

12. Three inductors in series, but placed too far apart for mutual inductance, have values of 50 μh, 25 μh, and 15 μh. What is the total inductance?

13. Three inductors in parallel have inductances of 20 μh, 25 μh, and 100 μh. What is the total inductance if there is no mutual inductance?

Answers

If your answers to the test questions do not agree with the ones given below, review the frames indicated in parentheses after each answer before you go on to the next chapter.

1. Any current has an associated magnetic field.   (1)

2. The associated magnetic field moves across (cuts) the conductor, inducting an emf.   (5)

3. It opposes any change in current flow.   (10-12)

4. (a) L; (b) henry; (c) h.   (13)

5. 20 microvolts   (18-20)

6. (1) number of turns in the coil, increase; (2) length of the coil, decrease;
   (3) cross-sectional area of the coil, increase; (4) permeability of the
   core, increase.   (21-28)

7. ⦙⦙⦙⦙⦙ (30)

8. (a) As the magnetic field expands, inductance produces a counter emf
   that opposes the buildup of current.   (44)
   (b) As the magnetic field collapses, inductance produces a counter emf
   that tends to prolong current flow.   (44-45)

9. 5 milliseconds   (48-49)

10. 2 microseconds   (54)

11. A voltage is induced in Coil B any time there is motion of the magnetic
    field associated with Coil A, whether the field is expanding or contract-
    ing.   (63)

12. 90 μh   (64)

13. 10 μh   (65)

CHAPTER NINE

Capacitance

In Chapter Eight, you learned that inductance opposes any change in current. Capacitance is the property of an electric circuit that resists, or opposes, any change of voltage in a circuit. That is, if applied voltage is increased, capacitance opposes the change and delays the voltage increase across the circuit. If applied voltage is decreased, capacitance tends to maintain the higher original voltage across the circuit, thus delaying the decrease. Consequently, the most noticeable effect of capacitance in a circuit is that voltage can neither increase nor decrease as rapidly in a capacitive circuit as it can in a circuit that does not include capacitance.

Capacitance is also defined as that property of a circuit that enables energy to be stored in an electric field. Unplanned, or "natural," capacitance exists in many electric circuits. In this book, however, we are concerned only with capacitance that is designed into the circuit by means of devices called capacitors.

When you have finished this chapter you will be able to:

- describe the construction of a capacitor;

- explain the factors affecting the value of capacitance;

- describe the charge and discharge of circuits that include capacitance and resistance;

- calculate RC time constants;

- calculate total, or equivalent, values of capacitance in series or in parallel;

- calculate the working voltage of a capacitor; and,

- describe some common types of capacitors.

The Capacitor

1. The <u>capacitor</u> is essentially a device that stores electrical energy. It can be charged and discharged, as we shall see later. The capacitor is a manufactured device that introduces capacitance to a circuit, just as the resistor presents resistance to current flow and the inductor greatly increases the inductance of a conductor. The capacitor is used in a number of ways in electrical circuits. It may block d-c in a-c portions of a circuit, since it is effectively a barrier to direct current (but not to alternating current). It may be part of a tuned circuit—one such application is in

the tuning of a radio to a particular station. It may be used to filter a-c out of a d-c circuit. Most of these are advanced applications that are beyond the scope of this book; however, a basic understanding of capacitance is necessary to the fundamentals of alternating current theory. Let's start with the construction of the capacitor.

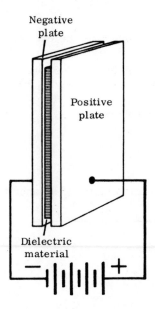

Figure 9-1. A simple capacitor.

Refer to Figure 9-1 for Frames 1 through 6.

A capacitor in its simplest form is shown connected to a battery in Figure 9-1. It consists of two metal plates, a _____ plate and a _____ plate, separated by a thin layer of insulation called the _____.

- - - - - - - - -

positive; negative; dielectric material   (Note: the dielectric material is usually called simply the "dielectric," pronounced "dye-electric." The prefix "di-" means "through" or "across.")

2. You learned early in this book that a body with an excess of electrons is negatively charged, while a body with a deficiency of electrons is positively charged. When a capacitor is connected across a voltage source, such as a battery, the voltage forces electrons onto one plate, making it

(negative/positive) _____.

- - - - - - - - -

negative

3. The other terminal of the battery pulls electrons off the other plate, mak-

ing it _____.

- - - - - - - - - -

positive

4. Electrons cannot flow through the dielectric, because it is an insulator.
   Since it takes a definite quantity of electrons to "fill up" (charge) a capacitor, it is said to have <u>capacity</u>. This characteristic is referred to as
   <u>capacitance</u>. A capacitor has a capacity for a certain quantity of what?

   _____

- - - - - - - - - -

   electrons

5. What do we call the characteristic of a capacitor that enables it to take

   a charge? _____

- - - - - - - - - -

   capacitance

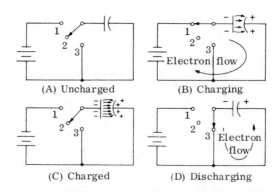

Figure 9-2.   Capacitor action.

Refer to Figure 9-2 for Frames 6 through 13.

6. The basic action of a capacitor is illustrated in four different states in
   Figure 9-2. Note that this figure introduces the symbol for a capacitor.
   The symbol is most clearly seen in View A. Draw the symbol for a cap-
   acitor.

- - - - - - - - - -

⊣⊢ (Note: The symbol for a capacitor is often seen as two straight lines rather than as a straight and a curved line.)

7. The three-position switch has one open and two closed positions. As shown in View A, the capacitor is in a balanced, or uncharged, condition. There is no way to place a charge on its plates because the battery is not in the circuit. The battery is connected into the circuit when the switch is

placed in what position? _____

- - - - - - - - - -

1

8. When the switch is set to position 1, as in View B, a surge of battery cur- rent immediately begins to charge the capacitor. There is only a momen- tary surge of electrons, after which direct current is completely stopped by the dielectric of the capacitor. Free electrons do not pass from one plate of the capacitor to the other. When we say that a capacitor has a charge (is unbalanced), do we mean that the charges on the two plates are

equal or unequal? _____

- - - - - - - - - -

The charges are unequal.

9. The natural tendency of the capacitor is to restore its own balance by dis- charging through the battery in a clockwise direction (View B). This ten- dency is small at first because the imbalance between the two plates is small, but the tendency to discharge becomes stronger as the capacitor is charging (that is, as the imbalance becomes greater).
   In your study of inductance, you learned that the action of a magnetic field induces a counter emf in a conductor. The tendency of a capacitor to discharge is also a counter emf. This tendency results from the dif- ference in potential between the two plates, which have different charges. Some current actually flows in the circuit as shown in View B, because the battery is moving electrons onto one plate and pulling them away from the other. When the counter emf of the capacitor has risen to equal the battery voltage, we say that the capacitor is charged. At this point, there is no more room for electrons on its negative plate, and the battery can- not pull any more electrons away from its positive plate. This condition is shown in View C. When the capacitor is fully charged, what happens

to the movement of electrons in the circuit? _____

- - - - - - - - - -

The movement of electrons ceases.

10. When the capacitor is fully charged (View C), the switch is returned to position 2. The capacitor is now a source of potential energy, somewhat as if it were a charged battery. Because of the unequal charges on its plates, there is a difference in potential between the negatively charged plate of the capacitor and terminal 3 of the switch. When the switch is set to position 3 (View D), a circuit is completed, and the charged capacitor is the source voltage. When the switch is set to position 3, what

happens in the circuit? _____

_____

- - - - - - - - - -

Current begins to flow; that is, the capacitor begins to discharge.

11. Current flows during the time the capacitor is discharging. The capacitor is completely discharged when the charges on its plates are again equal. What happens to the current in the circuit when the discharge is complete?

_____

- - - - - - - - - -

It stops flowing.

12. In your study of inductance, you learned that magnetic lines of force are associated with an inductor. Lines of force are also associated with a capacitor, and they are shown between the plates of the capacitor in Views B and C of Figure 9-2. These lines of force are not magnetic; they are

electrostatic, or simply electric. They originate on the _____

plate and terminate on the _____ plate.

- - - - - - - - - -

negative; positive

13. The lines of force are assumed to follow the same paths that negatively charged bodies (or electrons) would take if they were free to move. The lines of force would be present even if the plates of a capacitor were enclosed in a perfect vacuum. Electric lines of force cannot exist in a metallic conductor to any great extent, because equalizing currents are free to flow within the metal. What part of the capacitor prevents the flow of

equalizing currents? _____

- - - - - - - - - -

the dielectric (material)

Dielectric Materials

14. Various materials differ in their ability to support electric lines of force (flux); that is, to serve as dielectric materials for a capacitor. Materials are rated in their ability to support electric flux in terms of a number called a <u>dielectric constant</u>. The higher the value of the dielectric constant (other factors being equal), the better the material serves as a dielectric. Dry air is the standard by which other materials are rated. In inductors, the permeability of the core material was also related to that of dry air, which has a permeability (μ) of 1. Similarly, you should expect the dielectric constant of dry air to be _____.

- - - - - - - - - -

1

15. Some of the dielectric materials used in capacitors may surprise you. You have probably thought of air as a good insulator, and normally it is. But the plates of most capacitors are very closely spaced. If air were used in such capacitors, the current would arc (jump) across the plates. All of the materials in the list below are used as dielectrics in capacitors.

Which is the best dielectric? _____

Material	Dielectric-constant (Average values)
Air	1
Polystyrene........	2.5
Paraffin paper......	3.5
Mica	6
Flint glass........	9.9
Methyl alcohol	35
Glycerin..........	56.2
Pure water	81

- - - - - - - - -

pure water   (The key word is <u>pure</u>. Water capacitors are actually in use today in some high-energy applications, in which differences in potential are measured in thousands of volts.)

Unit of Capacitance

16. The unit for capacitance, C, is the farad (named for Michael Faraday), abbreviated f. The capacitance of a capacitor is proportional to the quantity of charge that can be stored in it for each volt difference in potential between its plates. A capacitor has a capacitance of 1 farad when a quantity of charge of 1 coulomb imparted to it raises its potential 1 volt. (One

coulomb is equal to 6.28×10^{18} electrons.) This relationship may be stated mathematically as

$$C = \frac{Q}{E}$$

where C is the capacitance in farads, Q is the quantity of charge in coulombs, and E is the difference in potential in volts. A capacitor of one farad would be of enormous dimensions, so two practical units are used instead: the microfarad (μf) and the picofarad (pf). The picofarad is one millionth of a microfarad, or 1×10^{-12} farads. (If you are a bit weak on negative powers of 10, 1×10^{-12} is the same as 0.000000000001, or a decimal point followed by 11 zeroes and a 1.) A microfarad (1×10^{-6} farads) corresponds to a microhenry in inductance (Chapter Eight). A capacitor of 1 μf has a capacitance of one- (thousandth/millionth)

_____ of a farad.

- - - - - - - - - -

millionth

Factors Affecting the Value of Capacitance

17. The capacitance of a capacitor depends on three factors:
1. The area of the plates
2. The distance between the plates
3. The dielectric constant of the material between the plates
   A greater plate area provides a greater capacity to store free electrons. When the area of the plates is increased, what happens to capacitance? _____

- - - - - - - - - -

It is increased.

18. Capacitance is directly proportional to the flux field between the plates. This field is stronger when the plates are more closely spaced. As the distance between the plates is increased, what happens to capacitance?

_____

- - - - - - - - - -

It is decreased.

19. Capacitance is directly related to the ability of the dielectric to prevent the movement of free electrons between the places. If the dielectric constant is increased, what happens to capacitance? _____

- - - - - - - - - -

It is increased.

20. Name the three factors affecting the value of capacitance, and state whether capacitance will be increased or decreased with an increase of each
factor.

    1. _____

    2. _____

    3. _____

- - - - - - - - - -

1. Capacitance is increased as the area of the plates is increased.
2. Capacitance is decreased as the distance between the plates is
increased.
3. Capacitance is increased as the dielectric constant is increased.

Charge and Discharge of a Resistive-Capacitive (R-C) Series Circuit

21. According to Ohm's Law, the voltage across a resistance is equal to the
current through it multiplied by the value of the resistance ($E = IR$). This
means that a voltage will be developed across a resistance only when

_____.

- - - - - - - - - -

current flows through it (or equivalent wording).

22. If no current flows through a resistor, what is the voltage drop across

the resistor? _____

- - - - - - - - - -

zero

23. View A of Figure 9-3 on the next page shows a voltage divider circuit
consisting of a resistor (R) and a capacitor (C). The source voltage (E_S)
is supplied by a battery. Switches S_1 and S_2 are "ganged" (indicated by
a dashed line) so that one switch is opened when the other is closed.
    View B shows the graphs for the source voltage (E_S), the charge cur-
rent (i_c), the difference in potential across the resistor (e_R) during
charge, and the difference in potential between the plates of the capacitor
(e_C) during charge. View C shows the graphs for the source voltage (E_S),
the discharge current (i_d), the difference in potential across the resistor
(e_R) during discharge, and the difference in potential between the plates
of the capacitor (e_C) during discharge. The time lines also serve as the zero lines
for the charge and discharge curves.

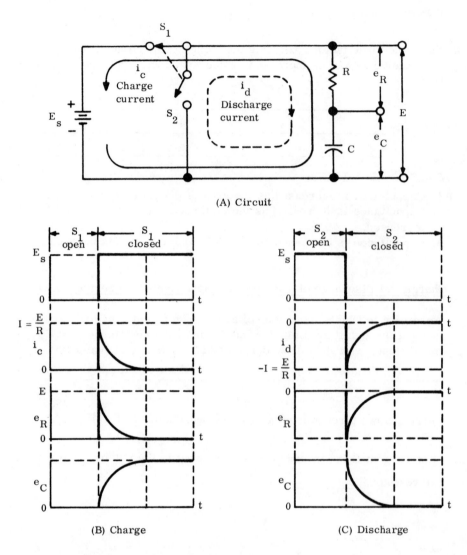

(A) Circuit

(B) Charge

(C) Discharge

Figure 9-3.   Charge and discharge of an R-C series circuit.

Refer to Figure 9-3 for Frames 23 through 39.

is because the time lines also serve as the zero lines for the charge and discharge curves. By displacing the zero lines, we can compare the to-tal increase or decrease for each value.

For the moment, look at View A only. When is the battery connected into the circuit? _____

_____

– – – – – – – –

when S_1 is closed (or, since the switches are ganged, when S_2 is open)

24. When S_1 is closed, electrons flow counterclockwise around the circuit containing the battery, the resistor, and the capacitor (View A). Remember that current does not flow <u>through</u> the capacitor, because its dielectric is an insulator. The current flow results from the fact that electrons accumulate on the negative plate of the capacitor and are pulled away from the positive plate as they are attracted by the positive terminal of the battery. Current flows momentarily through the resistor, but at first no difference in potential exists between the plates of the capacitor. At the instant current begins to flow, what is the relationship between the voltage

drop across the resistor and the battery (source) voltage? _____

_____

- - - - - - - - - -

They are equal (because there is no difference in potential between the plates of the capacitor, and the total difference in potential is across the resistor).

25. The electron flow stops when the capacitor is fully charged; that is, when the voltage across the capacitor (difference in potential between its plates) is equal to the battery voltage. Why is there now no current flow through

the resistor? _____

_____

- - - - - - - - - -

There is no difference in potential across the resistor; it is all across the capacitor.

26. There is a voltage drop across the resistor only as long as the capacitor

is (charging/discharging) _____.

- - - - - - - - - - -

charging

27. View B shows the division of the battery voltage E_s between the resistance and capacitance at all times during the charge process. Since the circuit is a voltage divider whose only components outside the battery are R and

C, the sum of e_R and e_C will at every instant be equal to _____.

- - - - - - - - - -

E_s (battery voltage)

28. View B shows graphs of the battery voltage E_s, the charge current i_c, and the voltage drops (differences in potential) across the resistor (e_R) and the capacitor (e_C). Remember that i_c is <u>not</u> the current through the capacitor; a capacitor is in effect an "open circuit" to direct current. (If this is confusing, review the explanation in Frame 24.) All three values are plotted against the same time. At the instant current begins to flow (S_1 closed), the entire source voltage E_s appears across

    (R/C) _____ and the voltage across C is _____.

    - - - - - - - - - -

    R; zero

29. The current flowing in the circuit soon charges the capacitor. Because the voltage on the capacitor is proportional to its charge, a voltage e_C will appear across the capacitor. This voltage opposes the battery voltage; that is, these two voltages "buck" each other. As a result, the voltage e_R across the resistor is $E_s - e_C$ (Kirchhoff's Law of Voltages). According to Ohm's Law, $e_R = i_c R$. Because E_s is fixed, as i_c decreases,

    what happens to e_C? _____

    - - - - - - - - - -

    The value of e_C increases.

30. The charging process continues until the capacitor is fully charged. At that point, $e_C = (E_s/e_R)$ _____.

    - - - - - - - - - -

    E_s

31. When the capacitor is fully charged, $e_R =$ _____.

    - - - - - - - - - -

    zero

32. View C shows the graphs of E_s, e_R, e_C, and i_d (discharge current). When the capacitor is fully charged and S_2 is open, all circuit values are as shown at the end of the charge cycle in View B. When S_2 is closed, the discharge cycle begins. At this point, what is the value of E_s in the circuit? _____

    - - - - - - - - - -

    Zero (because the battery is in the circuit only when S_1 is closed).

33. Since the discharge current i_d flows in a direction opposite to that of the charge current, its equation is shown as $-I = \dfrac{E}{R}$. At the instant S_2 is closed, what happens to the capacitor? _____

- - - - - - - - - -

It begins to discharge.

34. The graphs in Figure 9-3 show values between zero and maximum, with no specific magnitudes indicated. What is the value of e_C at the instant S_2 is closed? _____

- - - - - - - - - -

maximum

35. When S_2 is closed, the discharge current i_d is maximum and decreases to zero as the capacitor loses its charge. What happens to e_R during the discharge cycle? _____

- - - - - - - - - -

It decreases from maximum to zero.

36. When C begins to discharge, what happens to the charge current? _____

_____

- - - - - - - - - -

It begins to decrease.

37. When S_2 is closed, S_1 is opened and the battery is removed from the circuit. What provides the source voltage for current flow during the discharge cycle? _____

- - - - - - - - - -

the capacitor

38. During the discharge cycle, e_R is equal and opposite to _____.

- - - - - - - - - -

e_C

39. As the capacitor discharges, e_C decreases. Since e_C is the source voltage during discharge, i_d also decreases. In accordance with Ohm's Law, e_R also decreases. At the end of the discharge cycle, all values are

zero. The buildup of voltage (charge) on a capacitor and its decay (discharge) do not occur instantly, because the counter emf of a capacitor opposes any change of voltage in the circuit. The graphs in View C show that the voltages drop rapidly from their initial values when C begins to discharge and then approach zero more slowly. When i_d, e_R, and e_C all

reach zero, what is the state of the capacitor? _____

- - - - - - - - - - -

It is fully discharged.

The actual time it takes a capacitor to charge or discharge is important in advanced electricity and electronics. Since the charge or discharge time depends on the values of capacitance and resistance, a circuit can be designed for the proper timing of certain electrical events. We will study the RC time constant in the next section.

This is a convenient place to take a break.

RC Time Constant

40. Just as an L/R time constant was associated with inductance, an RC time constant expresses the charge and discharge times for a capacitor. You should recognize the values, since they correspond to the values for an R-L series circuit. The RC time constant is the time required to charge a capacitor to 63.2 percent of its maximum voltage or to discharge it to

_____ percent of its maximum voltage.

- - - - - - - - - -

36.8

41. The value of the time constant in seconds is equal to the product of the circuit resistance in ohms and its capacitance in farads.

$$R \text{ (ohms)} \times C \text{ (farads)} = t \text{ (seconds)}$$

This basic equation is not very useful, because capacitance is never given in farads. However, resistance is often in megohms (the prefix "meg-" means million), and capacitance is usually in microfarads or picofarads. Therefore, the following relations are useful in calculating the RC time constant.

$$R \text{ (megohms)} \times C \text{ (microfarads)} = t \text{ (seconds)}$$

$$R \text{ (ohms)} \times C \text{ (microfarads)} = t \text{ (microseconds)}$$

$$R \text{ (megohms)} \times C \text{ (picofarads)} = t \text{ (microseconds)}$$

One set of values illustrates this in Figure 9-4, on the following page. A circuit is shown with the values of R and C assigned. Look at the illustration, then answer the question that follows it.

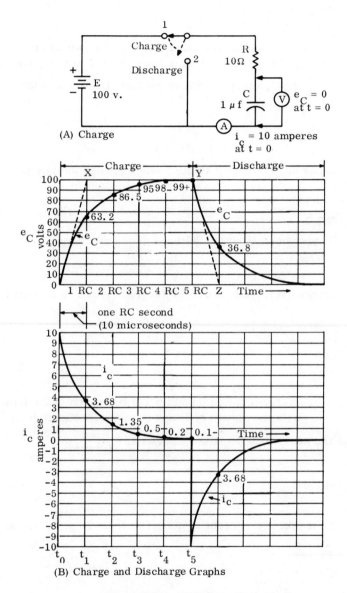

(A) Charge

(B) Charge and Discharge Graphs

Figure 9-4. RC time constant.

Refer to Figure 9-4 for Frames 41 through 49.

What units are shown for the following values?

resistance _____

capacitance _____

time _____

ohms; microfarads; microseconds

42. In Figure 9-4, the rise in the voltage e_C across the capacitor is plotted against a time line.   The increments of time in View B are shown in time constants (1 RC, 2 RC, etc.).   What is the time from 0 to 1 RC? _____

- - - - - - - - - -

10 microseconds

43. When e_C reaches 63.2 v., what is the voltage across the resistor?

_____

- - - - - - - - - -

36.8 v.   (Remember: The two voltages added must equal the source voltage, 100 v.)

44. During the charge cycle, the switch in the R-C circuit is in what position?

_____

- - - - - - - - - -

1

45. As the capacitor charges, the momentary current i_c from the battery rapidly decreases. With the values of R and C shown in Figure 9-4, how long does it take the current to decrease to 3.68 a. ? _____

- - - - - - - - - -

10 microseconds

46. How long does it take the current to decrease from 10 a. to 0.5 a. ?

_____

- - - - - - - - - -

30 microseconds (3 RC time constants or t_3)

47. If R in the circuit is 8Ω and C is 1.5 μf, how long will it take the voltage across the capacitor (e_C) to reach 63.2 v. ? _____

- - - - - - - - - -

12 microseconds (8 x 1.5)

48. $R = 10\Omega$ and $C = 2\,\mu f$. How long will it take the capacitor to charge to 98 percent of full charge? (Hint: Look at the graph in View A.) _____

- - - - - - - - - -

80 microseconds   (1 RC = 20 microseconds; 4 RC = 4 x 20)

49. The mathematics of the RC time constant is very simple. To visualize what actually happens in an RC circuit, however, you must remember that a capacitor charges by piling up free electrons on the negative plate, while free electrons are pulled away from the positive plate. If the capacitance is large, more electrons are accumulated before the capacitor is fully charged. During discharge, a large resistance presents a greater obstacle to the flow of free electrons, so it takes longer for the capacitor to discharge, or drain off the free electrons. Therefore, the RC time constant is greater if the capacitance is large (because there are more free electrons to drain off) or if the resistance is large (because a small current requires more time to drain off the electrons). Remember that a capacitor charges quickly at first and then more slowly reaches its maximum voltage, or difference in potential between its plates; it also discharges more quickly at first, and then the rate slows down. Define the RC time constant in terms of these charge and discharge character-istics. _____

_____

- - - - - - - - - -

The RC time constant is the time it takes a capacitor to charge to 63.2 percent of its maximum voltage or to discharge to 36.8 percent of its maximum voltage.

Capacitors in Parallel and in Series

50. Like resistors or inductors, capacitors may be connected in series, in parallel, or in a series-parallel combination. As we shall see, however, total capacitance is found in a different way. View A of Figure 9-5, which appears on the following page, shows three capacitors connected in parallel across a battery. If the voltage across C_1 is 100 v., what is the voltage across C_2? _____

- - - - - - - - - -

100 v.

51. We have seen that one factor affecting capacitance is the area of the plates. All the positive plates in Figure 9-5 are connected together and are electrically at the same point, and all the negative plates are also

electrically "tied" together. If all three capacitors are identical, the total plate area of the three capacitors is how many times larger than the plate area of a single capacitor? _____

- - - - - - - - - -

three

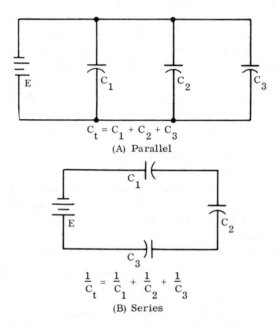

$$C_t = C_1 + C_2 + C_3$$

(A) Parallel

$$\frac{1}{C_t} = \frac{1}{C_1} + \frac{1}{C_2} + \frac{1}{C_3}$$

(B) Series

Figure 9-5.   Capacitors in parallel and in series.

Refer to Figure 9-5 for Frames 50 through 60

52. Since the three capacitors are identical, both plate spacing and dielectric are also identical. All three positive plates are connected, and all three negative plates are also connected. The three capacitors could be thought of as a single _____.

- - - - - - - - - -

capacitor

53. Since the three capacitors are identical, and their like plates are connected, they could be thought of as a single capacitor. If the capacitance of any one of the three capacitors is C, how might you express the capacitance of the combined capacitors? _____

- - - - - - - - - -

3C, or C + C + C

54. If the three capacitors are of different values, they could still be thought of as different parts of a single capacitor. In this case, total capacitance is the sum of the individual capacitances. Suppose the capacitors in Figure 9-5, View A, have the following values: 5 μf, 10 μf, and 15 μf. What is the total capacitance? _____

- - - - - - - - - -

30 μf

55. Look at the circuit in View B of Figure 9-5. The equation for total capacitance of capacitors in series (shown below the circuit) resembles the equation for total resistance of resistors in _____.

- - - - - - - - - -

parallel

56. In View B, C_1 = 5 μf, C_2 = 10 μf, and C_3 = 20 μf.

C_t = _____

- - - - - - - - - -

2. 86 μf   ($\frac{1}{C_t} = \frac{1}{5} + \frac{1}{10} + \frac{1}{20}$ = 0. 2 + 0. 1 +0. 05 = 0. 35; $C_t = \frac{1}{0.35}$ = 2. 86)

57. Two capacitors in series have values of 2 μf and 4 μf. What is the total capacitance? (Hint: You can use the product–over–sum method with capacitance, too.) _____

- - - - - - - - - -

1. 33 μf   ($\frac{2 \times 4}{6}$ = 1. 33)

58. Write the formula for total, or equivalent, series capacitance.

_____

- - - - - - - - - -

$$\frac{1}{C_t} = \frac{1}{C_1} + \frac{1}{C_2} + \frac{1}{C_3}$$

59. Write the formula for total, or equivalent, parallel capacitance.

_____

- - - - - - - - - -

$$C_t = C_1 + C_2 + C_3$$

60. You have been conditioned to calculate resistances in series by simple
addition and to use the more complicated formula for calculating parallel
resistance. This conditioning was reinforced by the fact that series and
parallel inductances are calculated in the same way. Therefore, it is
easy to forget that the rules are not the same for the calculation of total
capacitance. Make a general statement about the calculation of capacitance, both series and parallel, in comparison with the calculation of re-

sistance. _____

_____

- - - - - - - - - -

Parallel capacitance is calculated like series resistance, and series
capacitance is calculated like parallel resistance.

Voltage Rating of Capacitors

61. In selecting capacitors for an electrical circuit, you must decide not only
the values of capacitance but the amount of voltage to which the capacitors
will be subjected. As with any other electrical component, a capacitor is
designed to withstand a certain maximum voltage. If this voltage is exceeded, current will arc (jump) between the plates, despite the insulation
of the dielectric. This current will be great enough to damage the capacitor. The maximum voltage that can be steadily applied to a capacitor
without danger of arc-over is its working voltage. The working voltage
depends on the physical construction of the capacitor and is indicated by
the manufacturer. When a capacitor is selected, its working voltage
should be at least 50 percent greater than the maximum voltage it will
encounter in the circuit. (This additional 50 percent is a standard margin
of safety.) If a capacitor is expected to have a maximum of 100 v. applied

to it, its working voltage should be at least _____.

- - - - - - - - - -

150 v.

62. What may happen if the working voltage of a capacitor is too low?

_____

- - - - - - - - - -

The capacitor may be damaged.

63. A capacitor is to be used in a circuit in which it will be subjected to 500 v. Its minimum working voltage should be _____.

- - - - - - - - - -

750 v.

Types of Capacitors

64. There are two major groups of capacitors: <u>fixed</u> and <u>variable</u>. The names describe the difference between them. The fixed capacitor has a set value of capacitance that is determined by its construction. The construction of the variable capacitor allows a range of capacitances. Within this range, the desired value of capacitance is obtained by some mechanical means, such as by adjusting a screw or turning a shaft, as you do with the tuning knob on a radio. If you are designing a circuit in which some electrical effect is achieved by manually changing the capacitance, you

need to use at least one of which type of capacitor? _____

- - - - - - - - - -

variable

65. Capacitors come in a wide variety of sizes and constructions. Some capacitors used in extremely high-voltage applications may be 2 or 3 feet high, although most are so small that several may be comfortably held in one hand. Capacitors may be classified according to the type of material used as the dielectric, such as paper, oil, mica, and electrolyte. A common capacitor type is the <u>paper</u> capacitor, whose plates are strips of metal foil separated by waxed paper. The strips of foil and paper are rolled together to form a cylindrical cartridge, which is then sealed in wax to keep out moisture and to prevent corrosion and leakage. A metal lead is soldered to each plate and extends from the cylinder at either end. The illustration below shows the construction of a paper capacitor (A), the capacitor encased in cardboard (B), and a more rugged type of paper capacitor (C) that is hermetically sealed in a metal container.

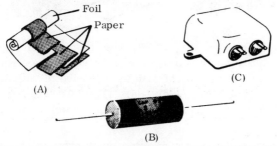

In a paper capacitor, what part of the capacitor is made of paper?

_____

- - - - - - - - - -

the dielectric

66. Oil capacitors are used at higher voltages, because a paper dielectric
might not provide sufficient insulation to prevent arcing across the plates.
These capacitors, whose dielectric is paper impregnated with oil, are
used mostly in radio and radar transmitters. Mica is a widely used dielectric material. Some typical mica capacitors are shown here.

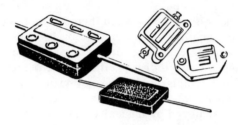

Name two types of capacitors that have better dielectrics than plain paper

capacitors. _____

- - - - - - - - - -

oil and mica

67. For capacitances greater than a few microfarads, the physical size of a
paper or mica capacitor becomes excessive. However, capacitors that
use an electrolyte as a dielectric may obtain large capacitances with small
physical dimensions. Unlike most other capacitors, electrolytic capacitors are marked with polarity (+ and −) and must be connected as indicated. Shown here are some typical electrolytic capacitors.

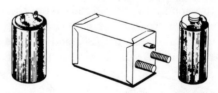

The electrolytic capacitor, widely used in electronic circuits, consists of
two metal plates separated by an electrolyte. The electrolyte, either
paste or liquid, is in contact with the negative terminal, and this combination forms the negative electrode. The dielectric is a very thin film of
oxide deposited on the positive electrode, which is an aluminum sheet.

In which of the following capacitors is polarity of connection important?

_____ electrolytic

_____ oil

_____ mica

_____ paper

- - - - - - - - -

only the electrolytic

68. So far we have discussed only fixed capacitors. Variable capacitors, as the name implies, may be adjusted to change the capacitance. A common application is the variable capacitor (A) in the drawing below. This type of capacitor is commonly seen in ordinary radios and is varied as stations are selected with the tuning knob. A screw-adjust may also be used to vary the capacitance, as shown in (B).

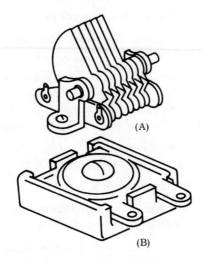

(A)

(B)

Since the plates of a variable capacitor must be free to move, a dielectric such as paper or mica cannot be used. Can you guess what is used as the dielectric of a variable capacitor? _____

- - - - - - - - - -

If you guessed air, you're right. (Air is not used as the dielectric of fixed capacitors because they would have to be too big, and because much better dielectrics can be used economically.)

Only the most common types of capacitors have been described. Others, such as the ceramic, are coming into use as the technology advances.
In this chapter you have learned how a capacitor is constructed as well

as some of the factors that affect capacitance. You have seen how capacitors are charged and discharged and how the charge and discharge
times are related to the RC time constant.

You have learned to calculate RC time constants, total capacitance in
series and in parallel, and the working voltage of a capacitor.

Finally, you have been introduced to some of the more common types
of capacitors.

When you feel that you understand the material in this chapter, go on
to the Self-Test.

Self-Test

The following questions will test your understanding of Chapter Nine. Write
your answers on a separate piece of paper and compare them with the answers
provided following the test.

1. Describe the basic components of any capacitor.

2. Draw the schematic symbol for a capacitor.

3. Electrons do not pass between the plates of a capacitor, yet current flows
    in the circuit as the capacitor is charged. How is this possible?

4. When the battery is removed from a circuit that includes a charged capa-
    citor, the capacitor will discharge if a complete circuit is provided for
    current flow. Since there is no battery, what causes current to flow dur-
    ing discharge, and when will the current cease to flow?

5. Polystyrene has a dielectric constant of 2.5, while that of paraffin paper
    is 3.5. Which material is the better dielectric?

6. One microfarad is what fraction of a farad?

7. Name the three factors that affect capacitance and state whether an <u>in-
    crease</u> in that factor will increase or decrease capacitance.

8. The capacitor in the circuit below will either charge or discharge, depend-
    ing on the position of the switch.

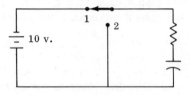

(a) Is position 1 of the switch the charge or discharge position?
(b) At the instant charging begins, what is the difference in potential
    across the resistor?
(c) At the instant charging begins, what is the difference in potential
    across the capacitor?

(d) At the instant the switch is moved to position 2 (assuming the capacitor is fully charged), what is the difference in potential across the capacitor?

(e) When the switch is in position 2, when will current stop flowing?

9. Define the RC time constant.

10. An RC circuit includes a 25-ohm resistor, a 5-microfarad capacitor, and a 100-volt battery. How long will it take the voltage across the capacitor to reach 63.2 v. ?

11. A 50-microfarad capacitor in series with a 3-kilohm resistor is connected to a 200-volt d-c source. What is the RC time constant of the circuit?

12. What is the total capacitance of two 50-microfarad capacitors connected in series?

13. What is the total capacitance of a series combination of two 100-microfarad capacitors connected in parallel with a 75-microfarad capacitor?

14. The highest voltage to be applied to a capacitor in a certain circuit is 300 v. What should be the minimum working voltage of the capacitor?

15. The design of a circuit requires a capacitor whose capacitance can be adjusted over a range of 10 to 100 microfarads. What general type of capacitor should be used in this circuit?

16. Fixed capacitors are classified according to the type of material used for the _____.

Answers

If your answers to the test questions do not agree with the ones given below, review the frames indicated in parentheses after each answer before you go on to the next chapter.

1. Any capacitor has two metal plates separated by a dielectric.   (1)

2.   (6)

3. Electrons, which are repelled by the negative terminal of the battery, accumulate on the negative plate of the capacitor, while electrons are attracted away from the positive plate by the positive terminal of the battery. (9)

4. There is a difference of potential between the plates of the capacitor, which will act as a battery. Current will cease to flow when the capacitor is discharged; that is, when the charges on the two plates are equal. (10-11)

5. Paraffin paper.   (15)

6. One millionth.   (16)

7. Area of the plates, increase; distance between the plates, decrease; dielectric constant, increase.   (20)

8. (a) Charge   (23-24)
    (b) 10 v.   (28)
    (c) Zero   (28)
    (d) 10 v.   (30)
    (e) When the capacitor is fully discharged.   (39)

9. The RC time constant is the time required to charge a capacitor to 63.2 percent of its maximum value or to discharge it to 36.8 percent of its maximum value.   (40)

10. 125 microseconds   (41-49)

11. 0.15 second (150,000 microseconds)   (41-49)

12. 25 μf   (50-60)

13. 125 μf   (50-60)

14. 450 v.   (61)

15. Variable capacitor.   (64)

16. dielectric   (65)

CHAPTER TEN

Inductive and Capacitive Reactance

In Chapters Eight and Nine, you learned that the inductance of a circuit acts to oppose any change of current flow in that circuit and that capacitance acts to oppose any change of voltage. These "reactions" are not important in direct current, because they are momentary and occur only when a circuit is first closed or opened. In alternating-current circuits, these effects become very important, because the direction of current flow is reversed many times each second; and the opposition presented by inductance and capacitance is, for practical purposes, constant.

In purely resistive circuits, either a-c or d-c, the term for opposition to current flow is resistance. When the effects of capacitance or inductance are present, as they often are in a-c circuits, the opposition to current flow is called reactance. The total opposition to current flow in circuits that have both resistance and reactance is called impedance.

When you have finished this chapter, you will be able to:

- calculate inductive reactance;

- calculate capacitive reactance;

- describe the phase relationships of resistive, inductive, and capacitive circuits; and,

- calculate impedance.

Calculation of Inductive Reactance

1. You learned in Chapter Eight that inductance generates a self-induced voltage that presents an opposition to the current already flowing in a conductor. This opposition to current flow is called inductive reactance. The amount of reactance depends on the magnitude of the self-induced voltage. For reasons that need not concern you here, this self-induced voltage, E_{ind}, depends on the frequency, f, of the alternating current, the amount of current, I, and a constant, 2π (2 pi). You may know that π represents the circumference of a circle divided by the diameter. It is normally rounded off to 3.14, so $2\pi = 6.28$. Naturally the inductance, L, is also a factor. From Ohm's Law, you know that $E = IR$. In an inductive

circuit (one that has inductance), E_{ind} is equal to current multiplied by inductive reactance (rather than resistance), because reactance is also an opposition to current flow. Therefore, $E_{ind} = 2\pi fLI$. (The values are switched around because the number usually comes first in a term such as $2\pi fLI$.) Thus, the self-induced voltage E_{ind} is the product of 2π and

what other three factors? _____

- - - - - - - - - -

f (frequency), I (amount of current), and L (inductance)

2. Compare the equations for E in a resistive circuit and E_{ind} in an inductive circuit:

<table>
<tr><td>Resistive</td><td>Inductive</td></tr>
<tr><td>E = IR</td><td>$E_{ind} = 2\pi fLI$</td></tr>
</table>

E corresponds to E_{ind}; I corresponds to I. R must correspond to what

is left in the inductive equation, or _____.

- - - - - - - - - -

$2\pi fL$

3. The opposition to current flow in an inductive circuit is <u>inductive reactance</u>, measured in ohms. Its mathematical symbol is X_L (pronounced "X sub L"). R represents resistance, while X represents reactance, whether it is inductive or capacitive. The type of reactance is identified by the subscript, so inductive reactance is X_L, and capacitive reactance (discussed later) is X_C.

Since R corresponds to X_L, the term X_L may be used in an equation for an inductive circuit in the same way that R is used in an equation for a resistive circuit. This does <u>not</u> mean that $R = X_L$. They are merely corresponding terms.

$$R = \frac{E}{I} \qquad\qquad X_L = \frac{E_{ind}}{I}$$

Since $E_{ind} = 2\pi fLI$, the term $2\pi fLI$ may be substituted for E_{ind} in any equation.

$$X_L = \frac{E_{ind}}{I} \qquad\qquad X_L = \frac{2\pi fLI}{I}$$

The I in the numerator and the I in the denominator of the second equation

cancel out, so $X_L = $ _____.

- - - - - - - - - -

$2\pi fL$ (Remember, X_L is not equal to R; it is merely the corresponding term.)

4. What is the equation for inductive reactance? _____

- - - - - - - - - -

$X_L = 2\pi fL$

5. What number is the constant 2π equal to? _____

- - - - - - - - - -

6.28

6. Since 2π is a constant equal to 6.28, we can find X_L if we know _____

and _____.

- - - - - - - - - -

frequency; inductance

7. In the equation for inductive reactance, X_L, the frequency is always in hertz (Hz), and the inductance is in henries (h). If L is in microhenries or millihenries, the value must be changed to henries in the equation. To convert millihenries to henries, multiply by _____.

- - - - - - - - - - -

0.001

8. The frequency of a circuit is 60 Hz and the inductance is 20 mh. What is X_L? _____

- - - - - - - - - -

7.536Ω   (X_L = 6.28 x 60 x 0.02 = 7.536   Note: If your answer was 1.2, you forgot to multiply by 2π.)

9. L = 10 h and f = 100 Hz.

$X_L =$ _____

- - - - - - - - - -

6280Ω

10. L = 4 mh and f = 200 Hz.

$X_L =$ _____

- - - - - - - - - -

5.024 Ω

11. L = 500 μh and f = 1500 Hz.

   X_L = _____

- - - - - - - - -

   4. 71 Ω (Note: If your decimal was in the wrong place, remember
   that you have to multiply μh by 0.000001 to get h.)

12. Other values may be found by applying Ohm's Law in the same way you
   do for resistive circuits. The alternating-current source voltage is 100 v.
   and the total inductive reactance in a purely <u>inductive</u> circuit is 50Ω. The

   total current is _____.

- - - - - - - - -

   2 a. $\left(I = \dfrac{E_{ind}}{X_L} \right)$

13. X_L = 40 Ω and I = 3 a.

   E = _____

- - - - - - - - - -

   120 v.   ($E_{ind} = IX_L$)

14. From now on, we will eliminate the subscript "ind" that identifies a voltage as resulting from the inductance in the circuit. Its purpose so far
   has been to remind you that self-induced voltage as well as source voltage
   must be taken into account in any circuit that includes inductance. As you
   will see later, a circuit is likely to include resistance, inductance, and
   capacitance. Therefore, we will be dealing with both resistance and reactance in the same circuit. The Ohm's Law equation for voltage in a
   resistive circuit is an old friend: E = IR. Its equivalent in an inductive
   circuit is E = IX_L. But you can't use Ohm's Law in an inductive circuit

   unless you know how to find X_L. What is the equation for X_L? _____

- - - - - - - - - -

   $X_L = 2\pi fL$

      As we mentioned earlier, reactance may be either inductive (X_L) or
   capacitive (X_C). We have seen that inductive reactance presents an opposition to current flow in any circuit that includes inductance. If a circuit has capacitance, there is also an opposition to current flow. It is
   called capacitive reactance, which we will study next.

      If you plan to stop pretty soon, this is a good place for a break.

Calculation of Capacitive Reactance

15. Capacitance, like inductance, presents a reactance, or opposition, to current flow. The basic symbol of reactance is X, and the subscript defines the type of reactance. In the symbol for inductive reactance, X_L, the subscript L refers to inductance. Following the same pattern, the symbol for <u>capacitive</u> reactance is _____.

- - - - - - - - -

X_C

16. The factors affecting capacitive reactance, X_C, are:

    the same constant, 2π,

    frequency, f, in Hz, and

    capacitance, C, in farads.

However, X_C is a little harder to calculate than X_L, because a reciprocal is involved:

$$X_C = \frac{1}{2\pi fC}$$

What is the capacitive reactance of a circuit operating at a frequency of 60 Hz, if the total capacitance is 133 μf? _____

- - - - - - - - -

20Ω (Note: If your answer was 0.050, you forgot to take the reciprocal. If the decimal was in the wrong place, remember that 133 must be multiplied by 0.000001.)

17. C = 50 μf and f = 100 Hz

$X_C =$ _____

- - - - - - - - -

31.85Ω

18. The alternating-current source voltage is 120 v. and the total capacitive reactance in a purely <u>capacitive</u> circuit is 40Ω. The total current is

_____. (Hint: Apply Ohm's Law, but use capacitive reactance instead of resistance.)

- - - - - - - - -

3 a. $\left(I = \dfrac{E}{X_C} \right)$

19. $X_C = 80\Omega$ and $I = 2$ a.

E = _____

- - - - - - - - -

160 v. $(E = IX_C)$

20. Of course, there is no such thing as a purely capacitive circuit in the "real" world, because circuits that include capacitors normally include resistors as well. (Even if they didn't, the conductors have resistance, which we usually ignore to simplify the problems in this book.) In fact, many circuits include resistance, inductance, and capacitance. In the next section, we will learn how to calculate the total impedance (that is, the total opposition to current flow) in a circuit that includes resistance, inductive reactance, and capacitive reactance. First, however, let's solve one more problem that involves both the calculation of total capacitance and application of the equation for capacitive reactance.

A circuit operating at a frequency of 100 Hz includes a parallel combination of a 10-microfarad and a 50-microfarad capacitor, and this parallel combination is connected in series with a 30-microfarad capacitor. What is the capacitive reactance of the circuit? (Give your answer to two decimal places.) _____

- - - - - - - - -

$X_C = 79.62\Omega$ Solution: First find the total capacitance, which is $20\,\mu f$. (Refer to Chapter Nine if you need to review this.) Then apply the equation for capacitive reactance.

$$X_C = \frac{1}{2\pi fC} = \frac{1}{6.28 \times 100 \times 20 \times 0.000001} = \frac{1}{0.01256} = 79.62$$

Phase Relationships of Resistive, Inductive, and Capacitive Circuits

21. In a purely resistive circuit, current rises and falls with voltage; it neither leads nor lags. Therefore, current and voltage are said to be in phase. Current and voltage are not in phase in inductive and capacitive circuits, because occurrences are not quite instantaneous in circuits that have either inductive or capacitive components. So it is time to introduce Eli the Iceman, who will help you to remember the phase relationships.

In the case of an inductor, voltage is first applied to the circuit, then the magnetic field begins to expand, and self-induction causes a "bucking" current to flow in the circuit, opposing the original circuit current. Voltage leads current (ELI) by 90 degrees. (ELI means that voltage, E, comes before current, I, in an inductive, L, circuit.)

When a circuit includes a capacitor, a charge current begins to flow and then a difference in potential appears between the plates of the capacitor. (For simplicity, we say that a voltage appears across the capacitor.) Current leads voltage (ICE) by 90 degrees. The phrase ELI the

ICE man will remind you that in an inductive circuit, voltage leads current by 90 degrees; and in a capacitive circuit, current leads voltage by 90 degrees.

In an inductive circuit, current (leads/lags) _____ voltage

by 90 degrees; in a capacitive circuit, current (leads/lags) _____ voltage by 90 degrees.

- - - - - - - - - -

lags; leads

22. In an inductive circuit, voltage leads current by _____ degrees. In a

capacitive circuit, current leads voltage by _____ degrees.

- - - - - - - - - -

90; 90

23. What is the phase relationship between voltage and current in an inductive

circuit ? _____

- - - - - - - - - -

Voltage leads current by 90 degrees.

24. What is the phase relationship between current and voltage in a capacitive

circuit ? _____

- - - - - - - - - -

Current leads voltage by 90 degrees.

25. What is the phase relationship between current and voltage in a resistive

circuit ? _____

- - - - - - - - - -

Current and voltage are in phase.

26. Disregard the resistance of the coil and connecting wires in View A of Figure 10-1 on the following page. View B shows sine waves for the source voltage E, the current I, and the voltage induced by the coil, E_{ind}. Therefore, you know that the source voltage is a-c. A small sine wave

is part of the symbol for an a-c voltage source. Draw the symbol. ____

- - - - - - - - - -

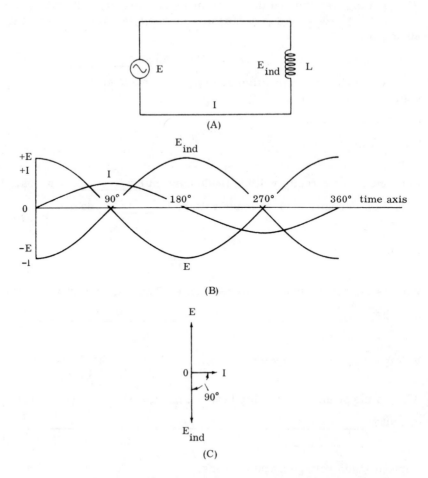

Figure 10-1. Sine waves of voltage, current, and induced voltage in an inductive circuit.

Refer to Figure 10-1 for Frames 26 through 35.

27. When an a-c voltage is applied across the circuit, which in this case includes only a coil, current flows through the coil and a magnetic field begins to expand, inducing a voltage in the coil. Remember from your study of the alternator (another name for an alternating-current generator) that the alternator itself has one or more coils (inductors). The armature cuts the greatest number of lines of force at zero degrees of rotation. Examine the labeled waveforms in View B. The waveforms show that at zero degrees, circuit voltage is maximum (positive/negative)

_____, while circuit current is _____.

- - - - - - - - - -

positive; zero

28. Circuit current reaches its maximum positive value 90 degrees later.
At this point, the armature is cutting zero lines of force, and applied

voltage (E) is (maximum/zero) _____.

- - - - - - - - - -

zero

29. Remember that the armature rotates counterclockwise, so the vectors
representing voltage and current also rotate counterclockwise. View C
shows the vectors for applied voltage (E), current (I), and the voltage induced in the coil (E_{ind}). Since voltage leads current in an inductive circuit, the applied voltage in View B reaches its maximum positive value

90 degrees (before/after) _____ the current.

- - - - - - - - - -

before

30. The vectors in View C show that E (leads/lags) _____ I by _____
degrees.

- - - - - - - - - -

leads; 90

31. The waveforms in View B show that the applied voltage (E) is at maximum
positive while the voltage induced in the coil (E_{ind}) is at maximum negative. The vectors for E and E_{ind} in View C show that E leads E_{ind} by

how many degrees? _____

- - - - - - - - - -

180 degrees

32. When E and E_{ind} are both at maximum (but of opposite polarity) I is at

_____.

- - - - - - - - - -

zero

33. The resultant of two vectors that are 180 degrees out of phase is always

zero. E and E_{ind} cancel out when both are maximum. Why? _____

_____

- - - - - - - - - -

because they are 180 degrees out of phase

34. What is the phase relationship between the current through the coil and the voltage induced in the coil? _____

_____

- - - - - - - - -

The current leads the induced voltage by 90 degrees. (If you had the wrong answer, see Frame 29.)

35. When we say that voltage leads current by 90 degrees in an inductive circuit, we are referring to (induced/applied) _____ voltage.

- - - - - - - - -

applied

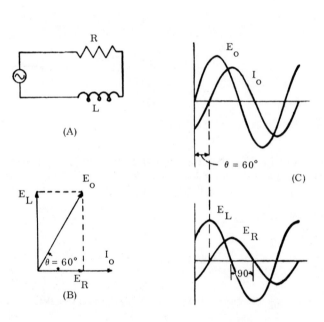

Figure 10-2. Resistance and inductance in series.

Refer to Figure 10-2 for Frames 36 through 39.

36. Because any practical inductor must be wound with wire that has resistance, it is not possible to obtain a coil without some resistance. For the purpose of calculation, this resistance may be considered as a separate resistor, R, in series with an inductor, L, as shown in View A of Figure 10-2. The resistance has been exaggerated in this example to clarify the explanation. In reality it would be comparatively small. The subscript "o", as used in Figure 10-2, means "output." It refers to the output of

the alternator. The alternating current, I_o, flows through both the resistor and the inductor, since they are in series. The voltage dropped across the resistor, E_R, is in phase with the current, but the voltage across the inductor, E_L, leads the current by 90 degrees. The vectors in View B show the relationships of the various phases, which are also represented by the waveforms in View C. What is the phase relationship between E_L and E_R? _____

- - - - - - - - - -

E_L leads E_R by 90 degrees.

37. What is the phase relationship between E_R and I_o? _____

_____

- - - - - - - - - -

E_R and I_o are in phase.

38. Angle θ represents the phase relationship between E_o and I_o. E_o leads I_o by how many degrees? _____

- - - - - - - - - -

60 degrees

39. The vectors E_L and E_R are two sides of a right triangle (a triangle that has a 90-degree angle) if the vector E_L is displaced to the position shown by the vertical dashed line in View B. The length of the hypotenuse (in this case, E_o) of a right triangle may be found if the lengths of the other two sides (here, E_L and E_R) are known. While the vectors represent various quantities as problems are solved, it is convenient to label the sides of the right triangle R and X and the hypotenuse Z. (This is easier than learning a different equation for each set of vectors.) In View B, let R = E_R, X = E_L, and Z = E_o. To keep track of the values, write in these labels on the triangle in View B. E_o is the resultant of the voltage drops across R and L. In a resistive circuit, these voltages are merely added; but when the voltages are out of phase, the total (or resultant) must be solved in some other way. One way is the use of the Pythagorean Theorem, which is mathematically stated as

$$Z = \sqrt{R^2 + X^2}$$

Let us assume that the voltage across the resistor, E_R, is 50 volts, and the voltage across the inductor, E_L, is 86.6 volts. We want to solve for the resultant voltage, E_o. The Pythagorean solution is

$$Z = \sqrt{50^2 + 86.6^2} = \sqrt{2500 + 7500} = \sqrt{10,000} = 100.$$

You can see immediately that 86.6^2 is not exactly 7500. It is actually 7499.56, but it is rounded off to 7500, which is close enough. If the specter of finding square roots by longhand (which you probably learned years ago and promptly forgot) bothers you, you should get a slide rule, a table of square roots, or an electronic calculator with a square root function. Anyone seriously interested in solving alternating-current problems should have one of these aids available, to simplify the extraction of square roots. Since you might not have such an aid available now, the problems in this book will involve numbers whose square roots are easy to extract by inspection.

Assign the following values to the vectors in Figure 10-2, View B:

$E_L = 44.72$ v.; $E_R = 40.0$ v. $E_o = $ _____

- - - - - - - - -

60 v. $(Z = \sqrt{40^2 + 44.72^2} = \sqrt{1600 + 2000} = \sqrt{3600} = 60)$

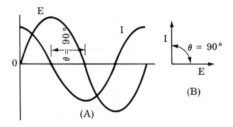

Figure 10-3. Phase relationship between
E and I in a capacitive circuit.

Refer to Figure 10-3 for Frames 40 through 43.

40. Figure 10-3 shows the waveforms (View A) and the vectors (View B) for voltage and current in a purely capacitive circuit. It is not at first obvious which leads, current or voltage (although you need to know to follow the discussion). You must pick a point where neither current nor voltage is at a specified level, such as maximum positive, zero, or maximum negative. Then you can see which waveform reaches that point first. For example, choose a point along the time line in View A (which starts at zero and is measured in degrees of rotation of an armature) where neither current nor voltage is at zero amplitude. The waveform that, starting from that point, reaches zero amplitude first is the leading waveform. In View A, which leads, E or I, and by how many degrees?

_____

- - - - - - - - -

I leads E by 90 degrees; that is, I reaches zero amplitude 90 degrees before E does.

41. View B of Figure 10-3 shows the vectors that represent the waveforms shown in View A.   Later in this book it will be necessary to know in which direction the vectors rotate.   Since you know that current leads voltage by 90 degrees in a capacitive circuit (which we have just reaffirmed), you also know that the vectors in View B rotate (clockwise/counterclockwise).

_____ .

- - - - - - - - - -

counterclockwise

42. Current leads the voltage across a capacitor by 90 degrees.   The connect-ing wires in a "purely" capacitive circuit have resistance, and current through a resistor is in phase with the voltage dropped across it.   If the voltage drop across the resistance was shown in View B, its vector would

be superimposed on which vector?   _____

- - - - - - - - - -

I

43. Voltage across a resistor (leads/lags) _____ voltage across a capacitor by 90 degrees.

- - - - - - - - - -

leads

44. In a circuit that includes a capacitor and a resistor, vectors can be drawn to show the phase relationship between E_C, the voltage across the capaci-tor, and E_R, the voltage across the resistor.   The vector for E_R is shown here.   Draw in the vector for E_C.

$$0 \xrightarrow{\quad E_R \quad}$$

- - - - - - - - -

$$0 \xrightarrow{\quad E_R \quad}$$
$$E_C \downarrow$$

45. Why was the E_C vector drawn pointing down?   _____

Why are the vectors at right angles?   _____

- - - - - - - - - -

because E_C lags E_R;   because the voltages are 90 degrees out of phase (or equivalent)

46. A circuit often includes resistance, inductance, and capacitance. E_L leads E_R, and E_C lags E_R, so all three vectors would look like this:

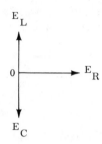

E_L and E_C are how many degrees out of phase? _____

- - - - - - - - -

180 degrees

47. When vectors are added, the new vector representing this vectorial sum is called the resultant. If E_L and E_C are equal, they cancel out, because they are 180 degrees out of phase. If E_o, the resultant, or vectorial sum, of all voltages were added to the vectors in Frame 46, its vector would be

superimposed on the vector for _____.

- - - - - - - - -

E_R (Since E_L and E_C are 180 degrees apart and thus cancel out, the only quantity left is E_R, which is also the resultant in this case.)

48. E_L and E_C are always 180 degrees out of phase, but they are likely to be unequal, as in this drawing.

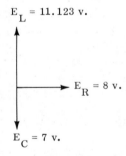

In this case, E_C cancels out part, but not all, of E_L. The resultant of E_L and E_C is _____ v.

- - - - - - - - -

4.123

49. E_C cancels out all but 4.123 v. of E_L, so the resultant voltages could be shown as:

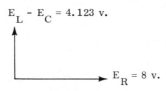

$$E_L - E_C = 4.123 \text{ v.}$$

$$E_R = 8 \text{ v.}$$

Use the Pythagorean Theorem, $Z = \sqrt{R^2 + X^2}$, to find the resultant voltage, E_O. _____

- - - - - - - - - -

$E_O = 9$ v.

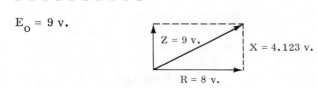

Z = 9 v.

X = 4.123 v.

R = 8 v.

(Note: You can make an approximate check of your answer in such problems if you remember that the value of Z, the hypotenuse of the right triangle, must always be greater than either of the other two sides, but less than their sum.)

50. $E_C = 16.32$ v., $E_L = 10$ v., and $E_R = 3$ v.

$E_O =$ _____

- - - - - - - - - -

$E_O = 7$ v. $(E_C - E_L = 6.32$ v.)

Earlier in this chapter, we called a circuit <u>inductive</u> if it included only inductance, and <u>capacitive</u> if it included only capacitance. These are the- oretical concepts, because any circuit also includes resistance. In the real world, a circuit often includes all three parameters: resistance, inductance, and capacitance. As we have seen, the effects of inductance and capacitance might not be equal. Since the two are always 180 degrees apart, one effect will cancel out <u>part</u> of the effect of the other. Thus, a circuit is:

- <u>capacitive</u> if the effect of capacitance outweighs the effect of inductance;
- <u>inductive</u> if the effect of inductance outweighs the effect of capacitance; and
- <u>resistive</u> if the effects of inductance and capacitance are equal, and therefore cancel out, leaving the effect of resistance only.

Since the phase relationships between voltage and current are always 90 degrees in capacitive or inductive circuits, their vectors form two sides of a right triangle, and the <u>resultant</u> forms the hypotenuse. The side of the right triangle representing resistance is labeled R, that representing reactance (whether inductive or capacitive) is labeled X, and the hypotenuse is labeled Z. Z is the mathematical symbol for impedance, or the total opposition to current flow in a circuit. We will study impedance next.

This is a good place for a break.

Impedance

51. We are more concerned with the <u>reactance</u> of capacitors and inductors than with the voltage drops across them. Since R represents resistance and X (either X_C or X_L) represents reactance, you can see why it is convenient to think of the vectors R, X, and Z. The resultant of X_L, X_C, and R can be found using the same equation as for E_L, E_C, and E_R. The term for <u>all</u> opposition to current flow in a circuit, resistive and reactive, is <u>impedance</u>, since both resistance and reactance impede current flow. From the terms we have been assigning to values in the Pythagorean equation, what do you think is the mathematical symbol for impedance?

_____

- - - - - - - - - -

Z

52. $Z = \sqrt{R^2 + X^2}$. If all the reactance in the circuit is inductive,

$Z = \sqrt{R^2 + \text{?}}$ _____

- - - - - - - - - -

$X_L{}^2$

53. If all the impedance in a circuit is capacitive, the equation for impedance

is _____.

- - - - - - - - - -

$Z = \sqrt{R^2 + X_C{}^2}$

54. The phase relationship between X_L and X_C is the same as that between E_L and E_C. Therefore, X_L (leads/lags) _____ X_C by

_____ degrees.

- - - - - - - - - -

leads; 180

55. If X represents the resultant (vectorial sum) of X_L and X_C, and X_L is greater than X_C, then $X = X_L - X_C$. Why? _____

_____

- - - - - - - - - -

Since X_L and X_C are 180 degrees out of phase, X_C cancels out part of X_L.

56. The net reactance of a circuit that includes both inductive and capacitive reactance is $X_L - X_C$. If X_C is larger than X_L, the net reactance is negative, but this makes no difference in the impedance equation, because the net reactance must be squared. X^2 is positive and $(-X)^2$ is also positive. The value $(X_L - X_C)^2$ is always (positive/negative) _____.

- - - - - - - - - -

positive

57. When the impedance of a circuit includes R, X_L, and X_C, both resistance and net reactance must be taken into account. Write the equation for impedance, and include both X_L and X_C in the equation. _____

- - - - - - - - - -

$$Z = \sqrt{R^2 + (X_L - X_C)^2}$$

58. $X_L = 400\Omega$, $X_C = 400\Omega$, and $R = 600\Omega$.

Z = _____

- - - - - - - - - -

600Ω

59. $X_L = 16\Omega$, $X_C = 10\Omega$, and $R = 8\Omega$.

Z = _____

- - - - - - - - - -

10Ω

60. $X_L = 30\Omega$, $X_C = 45\Omega$, and $R = 20\Omega$.

Z = _____

- - - - - - - - - -

25Ω

61. To review, the equation for X_L is _____ and the equation for X_C is _____ .

- - - - - - - - -

$$X_L = 2\pi fL; \quad X_C = \frac{1}{2\pi fC}$$

In this chapter you have learned that voltage and current are in phase in a purely resistive circuit. In an inductive circuit, however, voltage leads current by 90 degrees, while in a capacitive circuit, current leads voltage by 90 degrees. You have seen these relationships in the form of sine waves as well as vectors.

You have seen that the opposition to current flow in alternating-current circuits might be either of two kinds: resistance and reactance. Reactance, in turn, is either inductive (X_L) or capacitive (X_C), depending on whether the opposition to current flow is caused by inductance or capacitance.

The equation for inductive reactance includes frequency and a constant, 2π, as well as inductance. The equation for capacitive reactance includes these same terms in addition to capacitance.

You have learned to solve for impedance (Z) by the use of an equation, $Z = \sqrt{R^2 + X^2}$, which takes into account the phase relationships between current and voltage in capacitive and inductive circuits. When the opposition to current flow includes both X_L and X_C (which are 180 degrees out of phase) in addition to resistance, you must subtract the smaller from the larger value of reactance to arrive at X in the equation.

When you feel you understand the material covered in this chapter, turn to the Self-Test.

Self-Test

The following questions will test your understanding of Chapter Ten. Write your answers on a separate sheet of paper and check them with the answers provided following the test.

1. What are the factors affecting inductive reactance (in addition to the constant, 2π)?

2. What are the factors affecting capacitive reactance (in addition to the constant, 2π)?

3. A circuit whose frequency is 400 Hz has an inductance of 30 mh. X_L = ?

4. A circuit whose frequency is 100 Hz has a capacitance of 120 μf. X_C = ?

5. What is the phase relationship between voltage and current in a resistive circuit?

6. What is the phase relationship between voltage and current in an inductive circuit?

7. What is the phase relationship between voltage and current in a capacitive circuit?

8. What is the mathematical statement of the Pythagorean Theorem?

9. A circuit includes R, X_L, and X_C. The vector for R is shown below. Disregarding magnitude, complete the diagram by drawing the vectors for X_L and X_C.

$$0 \longrightarrow R$$

10. What is the term for all opposition to current flow? _____
    It includes both _____ and net _____ .

11. Write the equation for Z, including the terms R, X_L, and X_C.

12. $X_L = 27\Omega$, $X_C = 17\Omega$, and R = 24Ω. Z = ?

13. $X_L = 12\Omega$, $X_C = 28\Omega$, and R = 12Ω. Z = ?

14. What is the impedance of the circuit below?

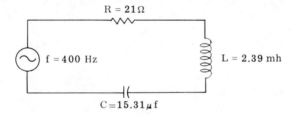

R = 21Ω

f = 400 Hz

L = 2.39 mh

C = 15.31 μf

Answers

If your answers to the test questions do not agree with the ones given below, review the frames indicated in parentheses after each answer before you go on to the next chapter.

1. inductance; frequency   (6)

2. capacitance; frequency   (16)

3. 75.36Ω   (4-11)

4. 13.27Ω   (16-17)

5. Voltage and current are in phase.   (21)

6. Voltage leads current (or current lags voltage) by 90 degrees.   (22)

7. Voltage lags current (or current leads voltage) by 90 degrees.   (22)

8. $Z = \sqrt{R^2 + X^2}$ (39)

9.

(46)

10. impedance; resistance; reactance   (51, 56)

11. $Z = \sqrt{R^2 + (X_L - X_C)^2}$ (57)

12. 26Ω (58-60)

13. 20Ω (58-60)

14. $Z = 29\Omega$ (58-61)

To check your work:

$X_L = 2\pi fL = 6.28 \times 400 \times 0.00239 = 6.00368 \text{ (or 6)}$

$X_C = \dfrac{1}{2\pi fC} = \dfrac{1}{6.28 \times 400 \times 0.0000\,1531} = 26.0019 \text{ (or 26)}$

$Z = \sqrt{R^2 + (X_L - X_C)^2} = \sqrt{21^2 + (6 - 26)^2} = \sqrt{441 + 400} = 29$

CHAPTER ELEVEN
Alternating-Current Circuit Theory

In previous chapters, the fundamental properties of resistive, inductive, and capacitive circuits were considered as isolated phenomena. The following phase relationships were seen to be true:

1. The voltage drop across a resistor is <u>in phase</u> with the current through it.
2. The voltage drop across an inductor <u>leads</u> the current through it by 90 degrees.
3. The voltage drop across a capacitor <u>lags</u> the current through it by 90 degrees.
4. The voltage drops across inductors and capacitors are <u>180 degrees out of phase</u>.

The solution of a-c problems is complicated by the fact that current varies with time as the a-c output of an alternator goes through a complete cycle. This is because the various voltage drops in the circuit vary in phase—they are not at their maximum or minimum values at the same time.

Alternating-current circuits frequently include all three circuit elements: resistance, inductance, and capacitance. In this chapter you will learn how the interaction of these elements affects the total circuit.

When you have finished this chapter, you will be able to:

● calculate voltage, current, and impedance in series L-C-R circuits;

● calculate voltage, current, and impedance in parallel L-C and parallel L-C-R circuits; and,

● calculate average and apparent power in a-c circuits.

The Series L-C-R Circuit

1. Figure 11-1 shows both the sine waveforms and the vectors for purely resistive, inductive, and capacitive circuits. The vectors indicate only direction, because the magnitudes are dependent on the values chosen for a given circuit. Remember that we are primarily interested in the <u>effective</u> (root-mean-square) values. If the individual resistances and reactances are known, Ohm's Law may be applied to find the voltage drops.

For example, we know that $E_R = IR$ and $E_L = IX_L$. According to Ohm's Law, $E_C =$ _____

– – – – – – – – – –

IX_C

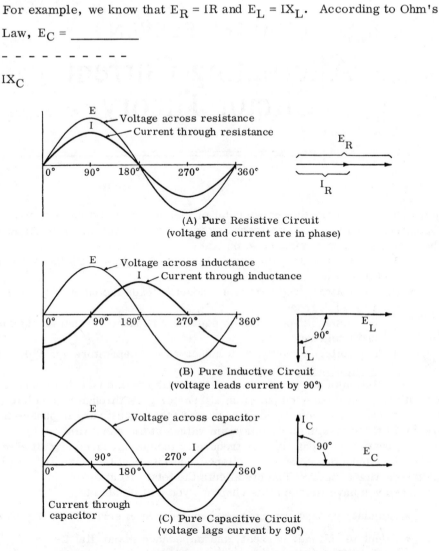

(A) Pure Resistive Circuit
(voltage and current are in phase)

(B) Pure Inductive Circuit
(voltage leads current by 90°)

(C) Pure Capacitive Circuit
(voltage lags current by 90°)

Figure 11-1. Graphic and vectorial representation
of R, L, and C circuits.

Refer to Figure 11-1 for Frames 1 through 5.

2. Current in an a-c circuit varies with time, so the voltage drops across
the various elements also vary with time. However, the same variation
is not always present in each at the same time (except in purely resistive
circuits) because current and voltage are not in phase. To review, in a
resistive circuit, the phase difference between voltage and current is

_____.

– – – – – – – – –

zero

3. In an inductive circuit, current (leads/lags) _____ voltage by

    _____ degrees.

- - - - - - - - - -

lags; 90

4. In a capacitive circuit, current (leads/lags) _____ voltage by

    _____ degrees.

- - - - - - - - - -

leads; 90

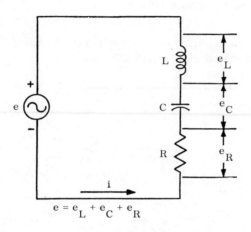

Figure 11-2. Voltage drops in an L-C-R circuit.

Refer to Figure 11-2 for Frames 5 through 7.

5. In practical applications, we are concerned mostly with effective values
    of current and voltage. To understand a-c theory, however, we need to
    know what happens from instant to instant.   Figure 11-2 shows the voltage
    drops across an inductor, a capacitor, and a resistor in a series a-c cir-
    cuit. Since phase relationships vary, Kirchhoff's Law of Voltages, as
    applied to a-c circuits, must be understood to mean that, at any instant,
    the sum of the voltage drops around a closed circuit is equal to the total
    applied voltage. The voltage drops e_L, e_C, and e_R are (effective/instan-

    taneous) _____ values.

- - - - - - - - - -

instantaneous

6. Since Kirchhoff's Law of Voltages <u>does</u> apply to instantaneous values in an a-c circuit, $e = e_R + e_L + e_C$. This relationship is true at any instant of time. But an instantaneous a-c voltage cannot be determined by Ohm's Law, which holds true only for maximum, effective, or average values, as in the equations in Frame 1. An a-c quantity, however, is fully determined when its <u>effective</u> value and phase (in respect to some standard, as we shall see later) are known. Therefore, the method of analysis by vectors, which show phase as direction and magnitude as effective value, can be used to add either sine voltages or sine currents. If the a-c voltages or currents are of sine waveform, we may add them by the use of

vectors if we know what two things? _____

- - - - - - - - - -

effective values and phase angles

7. Kirchhoff's Law of Voltages, as applied to the values shown in Figure 11-2, may be stated mathematically as:

$$e = \underline{\hspace{5cm}}$$

- - - - - - - - - -

$e_R + e_L + e_C$

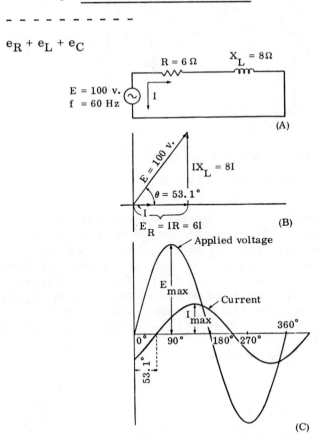

Figure 11-3. R-L circuit.

Refer to Figure 11-3 for Frames 8 through 11.

8. A 60-Hz, 100-volt alternator is connected to a circuit having a resistance of 6Ω in series with an inductive reactance of 8Ω, as shown in View A of Figure 11-3. The vector sum of the voltage drops across the resistance and the inductance is equal to the applied voltage (Kirchhoff's Law of Voltages). Since E_R is in phase with I, and E_L leads I by 90 degrees, their vector sum, 100 volts, is the hypotenuse of the right triangle shown in View B. Then:

$$E^2 = E_R{}^2 + E_L{}^2 .$$

The line current, I (the current resulting from the source voltage), may be found by substituting equivalents:

$$(100)^2 = (IR)^2 + (IX_L)^2$$
$$(100)^2 = 36I^2 + 64I^2$$
$$(100)^2 = 100I^2$$
$$I^2 = 100$$
$$I = 10 \text{ a.}$$

Let us look at a different 60-Hz alternator, using new values for R and L. $R = 3\Omega$ and $X_L = 4\Omega$. The applied voltage (the vector sum of E_R and E_L) is 25 volts. What is the line current? _____

- - - - - - - - - -

$I = 5$ a.

9. Since Ohm's Law may be applied to effective a-c values, I in Frame 8 could have been found by first finding Z and then solving for I. In a resistive circuit, $I = \dfrac{E}{R}$. In an a-c R-L circuit, what equation can be used? (Hint: Remember that impedance includes both resistance and reactance.)

_____

- - - - - - - - - -

$I = \dfrac{E}{Z}$

10. In an a-c R-L circuit, what is the equation for Z? _____

- - - - - - - - - -

$Z = \dfrac{E}{I}$

11. The angle θ (theta) in View B of Figure 11-3 indicates the phase relationship between line current and applied voltage. As shown in Views B and
C, current (leads/lags) _____ voltage by _____ degrees.

- - - - - - - - - -

lags; 53.1

12. After new values were substituted in Frame 8, the phase angle remained
53.1 degrees, because the ratio between the two short sides of the right
triangle was unchanged. (The ratio 8/6 is the same as the ratio 4/3.)
Trigonometry may be used to find the phase angle if the lengths of any
two sides of a right triangle are known. A table of sines, cosines, and
tangents is provided in Appendix II of this book. The <u>tangent</u> (abbreviated
"tan") of an angle in a right triangle is the ratio of the <u>opposite</u> side to the
<u>adjacent</u> side, relative to the angle. Here is the classic right triangle
with angle θ and the three sides indicated.

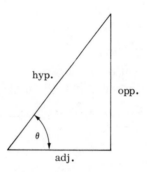

The adjacent side forms one side of the angle, the opposite side is opposite the angle, and the longest side is the hypotenuse. The tangent of
angle θ can be stated mathematically as:

$$\tan \theta = \frac{\text{opp.}}{\text{adj.}}$$

If the opposite side is 4 (volts, ohms, or any other value) and the adjacent
is 3, what is tan θ ? (State your answer to four decimal places.) _____

- - - - - - - - - - -

1. 3333

13. Look up the tangent, 1.3333, in the tables in Appendix II. The angle (to
the nearest 0.1 degree) of which 1.3333 is the tangent is _____
degrees.

- - - - - - - - - -

53.1   (Note: You found that the nearest tangent to 1.3333 shown in the tables is 1.3319; however, this is close enough for our purpose.)

14. The ratio of the opposite side to the hypotenuse is the <u>sine</u> of the angle:

$$\sin \theta = \frac{\text{opp.}}{\text{hyp.}}$$

If the opposite side is 4 and the hypotenuse is 5, what is $\sin \theta$? _____

- - - - - - - - - -

0.8000

15. The ratio of the adjacent side to the hypotenuse is the <u>cosine</u> of the angle:

$$\cos \theta = \frac{\text{adj.}}{\text{hyp.}}$$

If the adjacent side is 3 and the hypotenuse is 5, the $\cos \theta$ = _____.

- - - - - - - - - -

0.6000

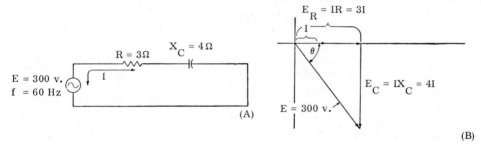
(A)   (B)

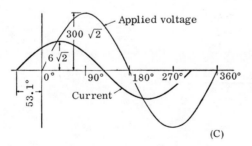
(C)

Figure 11-4.   R-C circuit.

Refer to Figure 11-4 for Frames 16 through 24.

16. The current, voltages, and phase angle of a series R-C circuit are found in the same way as for an R-L circuit. Figure 11-4, View A, shows the schematic of a circuit that has an a-c power supply, a resistor, and a

capacitor, with the values as shown in the schematic. As in the R-L circuit, the vector sum of the voltage drops across R and C equals the applied voltage E. The vectors are shown in View B. Since voltage across
a resistor is in phase with the current through it, the circuit is effectively

(resistive/capacitive) _____.

- - - - - - - - - -

capacitive

17. Can you explain why the vector for E_C in View B is drawn downward in-

stead of upward? _____

- - - - - - - - - -

Voltage lags current in a capacitive circuit.

18. Since E_R is in phase with I, and E_C lags I by 90 degrees, their vector
sum, 300 volts, is the hypotenuse of the right triangle shown in View B.
The line current can be calculated as in an R-L circuit:

$$E^2 = E_R{}^2 + E_C{}^2$$
$$(300)^2 = (3I)^2 + (4I)^2 = 25I^2$$
$$300 = 5I$$
$$I = \underline{\hspace{1cm}}$$

- - - - - - - - - -

60 a.

19. $E_R = IR = \underline{\hspace{1cm}}$

- - - - - - - - - -

180 v.

20. $E_C = IX_C = \underline{\hspace{1cm}}$

- - - - - - - - - -

240 v.

21. Remember that tan θ is the ratio of the opposite to the adjacent. What is

tan θ here? $\underline{\hspace{1cm}}$

- - - - - - - - - -

1.3333   (240 ÷ 180 = 1.3333)   (Note: Tan θ could be a much larger number. See Appendix II. Easy numbers were used here to simplify calculation.)

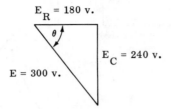

$E_R = 180$ v.

θ

$E_C = 240$ v.

$E = 300$ v.

22. The line current (leads/lags) _____ the applied voltage by

_____ degrees.

- - - - - - - - - -

leads; 53.1

23. What is the net reactance of the circuit? _____

- - - - - - - - - -

4 Ω   (There is no inductive reactance, so the net reactance is the same as X_C.)

24. Z = _____

- - - - - - - - - -

5 Ω To find Z, you had enough information to use any of these equations:

$$Z = \frac{E}{I} \qquad Z = \frac{X_C}{\sin \theta} \qquad Z = \sqrt{R^2 + X_C^2} \qquad Z = \frac{R}{\cos \theta}$$

25. When the three basic circuit elements of inductance, capacitance, and resistance are brought together in a single circuit, the voltage drops, current, and phase angle may be determined by combining the methods you have used for R-L and R-C circuits. View A of Figure 11-5 on the following page shows a series L-C-R circuit containing 6 Ω of resistance, 8 Ω of inductive reactance, and 16 Ω of capacitive reactance connected to a 60-cycle, 300-volt source. View B shows the sine waves and View C, the vector diagram for this circuit. Since current in a series circuit is everywhere the same, I is used as a reference for phase relationships.

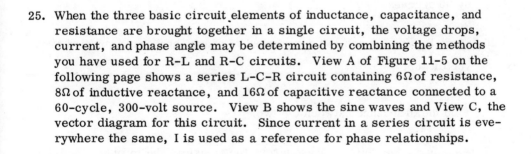

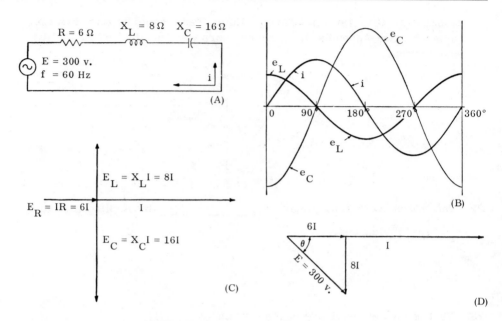

Figure 11-5. L-C-R circuit.

Refer to Figure 11-5 for Frames 25 through 33.

What is the phase relationship of each of these values with reference to I?

(1) E_R _____.

(2) E_L _____.

(3) E_C _____.

- - - - - - - - -

(1) E_R is in phase with I; (2) E_L leads I by 90 degrees; (3) E_C lags I by 90 degrees.

26. Since E_L and E_C are 180 degrees out of phase, how would you find their vector sum? _____

- - - - - - - - - -

Find the difference between the two.

27. As before, the voltage drops across the resistor, the inductor, and the capacitor are stated in terms of resistance or reactance so that the resultant line current can be found. Since E_C (16I) is larger than E_L (8I), the net reactive voltage (IX) is stated as (8I/16I/24I) _____.

- - - - - - - - - -

8I   (Note: It is only coincidence that the net reactive voltage has the same value as E_L.)

28. The net reactive voltage lags I. Can you explain why? _____

_____

- - - - - - - - - -

It has the direction of the larger voltage, which is E_C.

29. View D shows the resultant vector diagram. From the values given, solve for the line current, I. Reminder: $E^2 = (IR)^2 + (IX)^2$. _____

- - - - - - - - - -

I = 30 a.   Solution: $E = \sqrt{(6I)^2 + (8I)^2}$

$$= \sqrt{36I^2 + 64I^2}$$

$$= \sqrt{100I^2}$$

$$= 10I$$

$$I = \frac{E}{10} = \frac{300}{10} = 30$$

30. Now solve for E_R, E_L, and E_C, since you know I.

    (1) E_R = _____ v.

    (2) E_L = _____ v.

    (3) E_C = _____ v.

- - - - - - - - - -

(1) E_R = 180 v.   (6I = 6 x 30 = 180)
(2) E_L = 240 v.   (8I = 8 x 30 = 240)
(3) E_C = 480 v.   (16I = 16 x 30 = 480)

31. Z = _____

- - - - - - - - - -

10Ω $Z = \dfrac{E}{I} = \dfrac{300}{30} = 10$

32. Since Z is the hypotenuse of the impedance triangle, the side adjacent angle θ is R, or 6, and the side opposite angle θ is X, or 8 $(X_C - X_L)$,

tan θ is _____.

- - - - - - - - - -

1. 3333

33. The circuit is capacitive, because E_C is greater than E_L. What is the phase relationship between current and voltage? _____

_____

- - - - - - - - - -

Current leads voltage by 53.1 degrees.

34. Surprise! The phase angle is again 53.1 degrees. But don't get the idea that this is always true. The phase angle can be any value from zero to almost 90 degrees. (In reality, it is never quite 90 degrees because every circuit has some resistance.) The phase angle has been 53.1 degrees in our examples so far because we have, for convenience, used variations of the classic 3-4-5 triangle. That is, if one short side is 3 and the other is 4, the hypotenuse will be 5. Since R is always in phase with I, R will be the adjacent side of the phase angle. The other side, X (the resultant of X_L and X_C) will always lead or lag I by 90 degrees and will be the side opposite the phase angle.

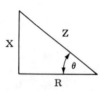
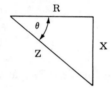

Use the table of trigonometric functions in Appendix II to solve for angle θ with the following information given. In each case, state whether I leads or lags E.

	R	X_L	X_C	tan θ	θ	I (leads/lags) E
(1)	5	10	8			
(2)	240	12	36			
(3)	10	200	300			
(4)	400	520	120			
(5)	13	12	64			

- - - - - - - - - -

(1) 0.4000; 21.8 degrees; lags
(2) 0.1000 (0.1016 is closest value in table); 5.8 degrees; leads
(3) 10.0000; 84.3 degrees; leads
(4) 1.000; 45.0 degrees; lags
(5) 4.0000; 75.9 degrees; leads

35. Let's examine further the five cases listed in Frame 34. In case 1, X_L is greater than X_C, so the net reactance (and thus, the circuit) is induc-

tive. In which other case is the circuit inductive? _____

- - - - - - - - - -

case 4

36. In which cases of Frame 34 is the circuit capacitive? _____

- - - - - - - - - -

cases 2, 3, and 5   (X_C is greater than X_L in these cases.)

37. Since X (net reactance) is one short side of the impedance triangle (whose hypotenuse is Z), and R (resistance) is the other short side, the phase angle depends on the relative values of net reactance (which may be either inductive or capacitive) and resistance. Let's consider another circuit that includes resistance, inductive reactance, and capacitive reactance.

What is the phase angle of the circuit if X_L and X_C are equal? _____

Why? _____

- - - - - - - - - -

The phase angle is zero, because the circuit is (in effect) purely resistive, and current and voltage are in phase. (Note: When X_L and X_C are equal, the net reactance is zero. In effect, the X side of the impedance "tri-angle" is zero, so the R side is equal to the hypotenuse. In other words, R = Z, or the impedance of the circuit is resistance only.)

38. The series L-C-R circuit illustrates three important points:
    ● The current in a series L-C-R circuit either leads or lags the applied voltage, depending on whether X_C is greater or less than X_L.
    ● A capacitive voltage drop in a series circuit always subtracts directly from an inductive voltage drop.
    ● The voltage across a single reactive element in a series circuit can have a greater effective value than that of the applied voltage. The circuit and impedance triangle below illustrate this point.

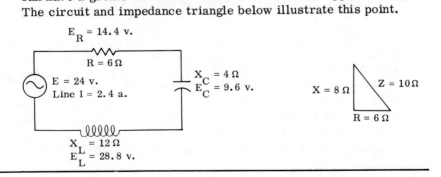

$$I = \frac{E}{Z} = \frac{24}{10} = 2.4 \text{ a.}$$

$E_R = IR = 2.4 \times 6 = 14.4 \text{ v.}$

$E_C = IX_C = 2.4 \times 4 = 9.6 \text{ v.}$

$E_L = IX_L = 2.4 \times 12 = 28.8 \text{ v.}$

E_L is greater than E.

   Can a voltage drop across a single element in a d-c circuit ever be greater than the applied voltage? Why or why not? _____

_____

- - - - - - - - - -

No, because according to Kirchhoff's Law of Voltages, the applied voltage is equal to the sum of the voltage drops in the circuit.

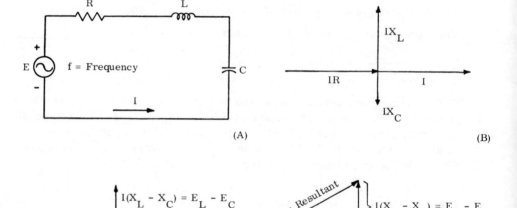

Figure 11-6.   Phase relationships in an L-C-R circuit.

Refer to Figure 11-6 for Frames 39 through 47.

39. You have now learned enough to calculate the current, impedance, voltage drops, and phase angle of any series L-C-R circuit. View A of Figure 11-6 shows a schematic diagram of a series circuit containing an inductance L, a capacitance C, and a resistance R. The circuit is connected to an a-c voltage source of magnitude E and frequency f. View B shows the initial vector diagram. The voltage drop across the resistance, IR, is drawn in phase with the current I. The voltage drop across the inductance is drawn above the zero axis, and the voltage drop across the

capacitance is drawn below the axis. View C shows the resultant vectors, and View D shows the voltage triangle with base E_R and vertical side $E_L - E_C$. The resultant voltage E, equal to the applied voltage, is the hypotenuse of the right triangle. You have solved for the phase angle by applying values of reactance and resistance in an impedance triangle. Since voltage and impedance are all related, any triangle formed from equivalent values of voltage will have the same ratio of sides as the impedance triangle. You saw an example of this in Frame 38. The impedance triangle shown there is a 3-4-5 triangle, since $R = 6\Omega$, $X = 8\Omega$, and $Z = 10\Omega$. If you were to label the R side of the triangle E_R, the X side E_X (which is $E_L - E_C$), and the Z side E (line voltage), the values of the sides would be 14.4 v., 19.2 v., and 24 v. Divide each value by the line current, 2.4 a., and you will see that you again have a 3-4-5 triangle.

You found tan θ and then phase angle θ by using values of impedance.

You can also find tan θ using values of _____.

- - - - - - - - - -

voltage

40. Can you find tan θ using values of current? Why or why not? _____

_____

- - - - - - - - - -

No; current is the same everywhere in a series circuit.

41. Let us assign some values for the circuit in Figure 11-6. $E_C = 40$ v., $E_L = 120$ v., $E_R = 40$ v., and the applied voltage $E = 89.44$ v. To find tan θ (which you need to find the phase angle), you need to know all of

these values except _____.

- - - - - - - - - -

E

42. If the circuit has the values assigned, what is the phase relationship between the current and the applied voltage? (Give the phase angle as part

of your answer.) _____

- - - - - - - - - -

The current lags the voltage by 63.4 degrees.

$$\text{Solution: } \tan \theta = \frac{E_L - E_C}{E_R} = 2.0000$$

$$\theta = 63.4$$

The circuit is inductive, so current lags voltage.

43. If E_C were 120 v. and E_L 40 v. , what would be the phase relationship between current and voltage ? _____

- - - - - - - - -

The phase angle would be the same, 63.4 degrees, but current would lead voltage.

44. Using another trigonometric function, the <u>cosine</u> of θ is equal to E_R/E (adjacent over hypotenuse), or:

$$\cos \theta = \frac{R}{Z}$$

$$Z = \frac{R}{\cos \theta}$$

If the phase angle is known, you can find the impedance of a circuit without knowing the reactance. For example, if the phase angle is 72.5 degrees and the resistance is 90 ohms, you can look in the tables for cos 72.5, which happens to be 0.3007. Working the equation, you find that $Z = 299.30\,\Omega$. The equation $\cos \theta = \frac{R}{Z}$ could be used to select circuit values that would result in a desired phase angle. What total impedance is required to produce a phase angle of 18.2 degrees if the circuit resistance is $950\,\Omega$? _____

- - - - - - - - -

$Z = 1000\Omega$ (Find $\cos \theta$, which is 0.9500. 950 ÷ 0.9500 = 1000.)

45. If X_L is greater than X_C, will the circuit be inductive or capacitive ?
_____ Why ? _____

- - - - - - - - - -

The circuit will be inductive because E_L is greater than E_C.

46. If X_L is less than X_C, will the circuit be inductive or capacitive ?
_____ Why ? _____

- - - - - - - - - -

The circuit will be inductive because E_L is greater than E_C.

47. If X_L and X_C are equal, the circuit will be _____.

- - - - - - - - - -

resistive

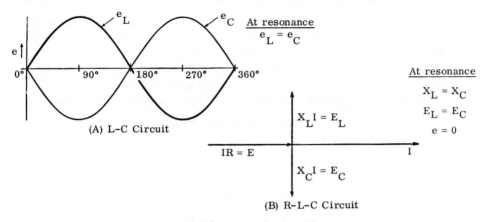

Figure 11-7. Sine waves and vectors for a series resonant circuit.

Refer to Figure 11-7 for Frames 48 through 51.

48. When X_L and X_C are equal, an interesting phenomenon called <u>resonance</u> occurs. (This is very important in electronics.) Figure 11-7 shows the voltage sine waves for e_L and e_C. Remember that e represents an instantaneous voltage while E represents an effective (rms) value. View A shows a circuit at resonance. From the sine waves shown, compare the voltage drops (both magnitude and polarity) across the inductor and the

capacitor <u>at any instant of time.</u> _____

_____

– – – – – – – – – –

The voltage drops will be equal in magnitude and of opposite polarity.

49. The voltage drops across the inductor and capacitor could be very high, but they cancel, so the source "sees" no reactive voltage. Therefore, the current in the circuit is limited only by the resistance. In a resonant cir-

cuit, impedance is (maximum/minimum) _____ and

line current is (maximum/minimum) _____.

– – – – – – – – – –

minimum; maximum

50. Since line current is maximum in a resonant circuit, the voltage drops across the inductor and capacitor are (maximum/minimum)

_____.

– – – – – – – – – –

maximum

51. Since both inductive reactance and capacitive reactance are dependent on the circuit frequency, the frequency at which resonance occurs can be worked out mathematically. Every circuit is resonant at some frequency. Resonance may be achieved by adjusting either L or C (or both) until the reactances are equal. This is what happens when you tune your radio to a given broadcast frequency. When inductive reactance and capacitive

reactance are equal, the circuit is said to be at _____.

- -. - - - - - - - - -

resonance

The Parallel L-C Circuit

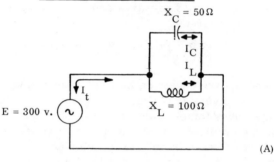

(A)

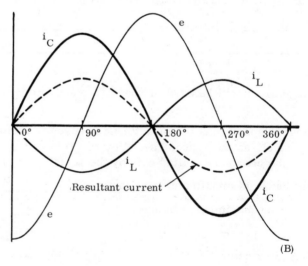

(B)

(C)

Figure 11-8. A parallel L-C circuit.

Refer to Figure 11-8 for Frames 52 through 59.

52. The schematic diagram in View A of Figure 11-8 shows an inductor and a capacitor in parallel. Assume that all the connecting wires are perfect conductors and thus have no resistance. The voltage across the inductor is 300 v. What is the voltage across the capacitor? _____

- - - - - - - - - -

300 v.

53. The source voltage, which is applied across both branches of the parallel L-C circuit, is used as a reference in the vector diagram in View C. The instantaneous values of capacitor current i_C, inductor current i_L, and source voltage e are shown as sine waves in View B. As in d-c parallel circuits, I_t is divided between the branches. Unlike d-c, I_t is not the arithmetical sum of the branch currents; it is the resultant (or vectorial sum).

The branch currents may be found by applying Ohm's Law. I_C is
_____ a., and I_L is _____ a.

- - - - - - - - - -

6; 3   Solution: $I_C = \dfrac{E}{X_C} = \dfrac{300}{50} = 6$ a.

$I_L = \dfrac{E}{X_L} = \dfrac{300}{100} = 3$ a.

54. As the vector diagram in View C indicates, I_C leads E by 90 degrees, while I_L lags E by 90 degrees. Why? _____

_____

- - - - - - - - - -

Current through any capacitor leads the voltage across it, while current through any inductor lags the voltage across it. (Remember ELI the ICE man.)

55. What is the phase relationship between I_C and I_L? _____

_____

- - - - - - - - - -

They are 180 degrees out of phase.

56. I_t, the resultant of I_C and I_L, is _____ a.

- - - - - - - - - -

3

57. The effect of the inductance is canceled out by the greater effect of the capacitance. The circuit appears to the source as (inductive/capacitive)

_____ .

- - - - - - - - - -

capacitive

58. It is important to keep track of the signs of the reactances in solving impedance problems, as we shall see. By convention, inductive reactance is regarded as having a plus sign, and capacitive reactance is regarded as having a minus sign. The total impedance, Z_t, of a two-branch parallel circuit may be found by the "product over sum" method you have used for solving parallel resistance. The equation for parallel resistance is shown below with the corresponding equation for impedance of a parallel L–C circuit.

$$R_t = \frac{R_1 \cdot R_2}{R_1 + R_2} \qquad Z_t = \frac{X_L \cdot X_C}{X_L + X_C}$$

But note that you must apply the signs of the reactances. If you merely inserted the values of X_L and X_C in the equation, without regard for the signs, you would get the wrong answer for impedance. Here are the right and wrong solutions for Z_t in the circuit shown in View A of Figure 11–8.

Wrong	Right
$Z_t = \dfrac{X_L \cdot X_C}{X_L + X_C}$	$Z_t = \dfrac{X_L \cdot X_C}{X_L + X_C}$
$= \dfrac{100 \cdot 50}{100 + 50}$	$= \dfrac{100 \cdot (-50)}{100 - 50}$
$= \dfrac{5000}{150}$	$= \dfrac{-5000}{50}$
$= 33.33\,\Omega$	$= -100\,\Omega$

The correct impedance is $-100\,\Omega$. The minus sign does not mean that the impedance is capacitive. It means that I_t is capacitive. This is true because X_C is less than X_L, so more current flows through the capacitor than through the inductor. Why is I_C greater than I_L? _____

_____

- - - - - - - - - -

X_C in this circuit offers less opposition to current flow than X_L.

59. Now try solving for total impedance in another L-C circuit that has a 50-ohm inductive reactance in parallel with a 200-ohm capacitive reactance.

What is Z_t? _____ . Is I_t inductive or capacitive? _____

- - - - - - - - - -

$Z_t = 66.67\Omega$; I_t is inductive. Solution: $Z_t = \dfrac{X_L \cdot X_C}{X_L + X_C}$

$$= \dfrac{50 \cdot (-200)}{50 - 200}$$

$$= \dfrac{-10,000}{-150}$$

$$= 66.67\Omega$$

60. You know that the total resistance of parallel resistors is less than the resistance of any branch. But this is not necessarily true when both X_C and X_L are present in a parallel circuit. For example, X_L and Z_t in Figure 11-8 are both 100Ω. Z_t can actually be much greater than either branch impedance. At resonance, X_L and X_C are equal in value (but opposite in effect). If you will examine our equation for Z_t for parallel branches, you will see that the denominator is zero, so Z_t is theoretically infinite. (In reality, it can only <u>approach</u> infinity, because there is always some resistance in the circuit.) So you can see that at resonance,

a parallel L-C circuit offers (maximum/minimum) _____ impedance.

- - - - - - - - - -

maximum

61. At resonance, line current in a parallel L-C circuit is (maximum/minimum) _____.

- - - - - - - - - -

It is minimum, because of Ohm's Law: $I_t = \dfrac{E}{Z}$.

62. Suppose X_L in a parallel L-C circuit at resonance is 200Ω. $X_C =$ _____

- - - - - - - - - -

200Ω

63. At resonance, the line current in a given L-C parallel circuit is 10 a. If either X_L or X_C is changed, line current will (increase/decrease)

_____.

- - - - - - - - -

It will increase because line current is minimum at resonance. If either X_L or X_C is changed, the circuit will not be resonant, Z_t will decrease, and I_t will increase.

64. Assume that in a parallel L–C circuit, $X_L = 30\,\Omega$, $X_C = 60\,\Omega$, and source voltage E = 100 v. $Z_t =$ _____

- - - - - - - - - -

$60\,\Omega$ (If your answer was $20\,\Omega$, you probably forgot to assign a minus sign to X_C.)

65. What is the line current in that circuit? _____

- - - - - - - - - -

1. 67 a.

66. Assume that in a parallel L–C circuit, $X_L = 180\,\Omega$, $X_C = 80\,\Omega$, and line voltage is 200 v. $Z_t =$ _____ . $I_t =$ _____ .

- - - - - - - - - -

$144\,\Omega$; 1.39 a.

67. Is Z_t in Frame 66 effectively inductive or capacitive? _____

- - - - - - - - - -

inductive

The Parallel L–C–R Circuit

68. The parallel L–C–R circuit may now be seen as essentially a parallel L–C circuit (subject to the interaction of L and C) with an added resistance in parallel with this circuit. View A of Figure 11–9 shows this circuit. At any nonresonant point (X_L not equal to X_C), what will be the phase relationship of the currents through L and C? _____

_____

- - - - - - - - - -

They will be 180 degrees out of phase.

69. The resultant current will be predominantly inductive or capacitive, depending on the branch offering the least impedance (opposition to) current flow. If X_L is greater than X_C, the impedance is effectively inductive,

and the line current is effectively _____.

- - - - - - - - - -

capacitive

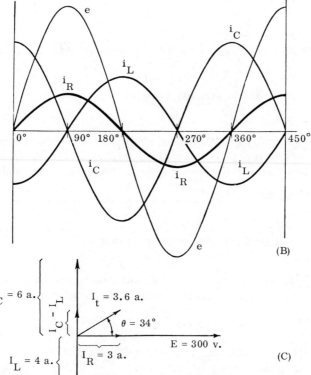

(A)

(B)

(C)

Figure 11-9.   Parallel L-C-R circuit.

Refer to Figure 11-9 for Frames 68 through 73.

70. An equivalent schematic diagram of this circuit would show either a capacitance or an inductance in parallel with the resistance. The resultant
line current is the vector sum of the reactive current and the resistive
current, as though the circuit comprised only a single reactance and a
resistance. I_t, then, as shown in View C, is the resultant of the resistive
current and the reactive current. What are the values of the resistive

current and the reactive current? _____

- - - - - - - - - -

The resistive current is 3 a. The reactive current is $I_C - I_L$, or 2 a.

71. Since we have, in effect, only two currents, I_R and I_X (reactive current), and they act at right angles (90 degrees) to one another, we can solve for I_t by applying our equation from the Pythagorean Theorem:

$$I_t = \sqrt{I_R^2 + I_X^2}$$

$$I_t = \underline{\hspace{1.5cm}}$$

- - - - - - - - -

3.6 a.

72. Since we now know I_t and E, we can find Z_t. $Z_t = \underline{\hspace{1.5cm}}$

- - - - - - - - -

83 Ω   (300 ÷ 3.6 is actually 83.33.) $Z_t = \dfrac{E}{I_t}$

73. Since $\cos \theta = Z/R$, or 0.83, we could have found the phase angle in the trigonometric tables in Appendix II (although it is shown in Figure 11–9). To the nearest whole degree, what is the phase relationship between line current and line voltage? _____

- - - - - - - - -

Line current leads line voltage by 34 degrees.

74. At the point of resonance, the two reactive currents effectively cancel out, leaving only the current through the resistance. At resonance, Z_t is the same as _____.

- - - - - - - - -

R

75. At resonance in any L–C–R parallel circuit, $\cos \theta = Z/R$, or 1. What is the value of angle θ ? _____

- - - - - - - - -

zero

76. At resonance, I_C and I_L are equal and cancel out, effectively leaving only the resistive current. Therefore, the phase angle of any resonant L–C–R circuit is zero. Why? _____

- - - - - - - - -

Current and voltage are in phase in a resistive circuit.

77. The chief characteristic of a parallel L–C circuit at resonance is its very high impedance. When resistance is added in parallel, however, this characteristic is lost, because the line current is limited only by the resistance. First consider an L–C circuit in which X_L and X_C are equal, or at resonance. Disregarding line resistance,

$$Z_t = \frac{X_L X_C}{X_L + X_C}$$

Since a minus sign must be carried with X_C, you can see that, regardless of the values, Z_t will be infinity, since it is equal to some value divided by zero. Assume that X_L and X_C are each $10\,\Omega$. Z_t approaches infinity. Now add a 10-ohm resistor in parallel and apply the other equation,

$$\frac{1}{Z_t} = \frac{1}{X_L} + \frac{1}{X_C} + \frac{1}{R}$$

$$= \frac{1}{10} + \frac{1}{-10} + \frac{1}{10}$$

$$= .1 + (-.1) + .1$$

$$\frac{1}{Z_t} = 0.1$$

$$Z_t = 10$$

Can you make a general statement about the effective relationship between total impedance and resistance when a resistor is added in parallel with a parallel L–C circuit at resonance? _____

_____

- - - - - - - - - -

You should have said something like: Total impedance is effectively equal to resistance.

If you plan to stop pretty soon, take a break now.

Power in A–C Circuits

78. In circuits that have only resistance, but no reactance, the amount of power absorbed in the circuit is easily calculated by Joule's Law: $P = I^2 R$. In dealing with circuits that include inductance or capacitance (or both), which is often the case in a–c electricity, the calculation of power is a more complicated process.

As we saw in Chapter Three, power is a measure of the rate at which work is done. The "work" of a resistor is to limit current flow. In doing this, the resistor dissipates heat, and we say that power is consumed or absorbed by the resistor.

Capacitors and inductors also oppose current flow, but they do so by producing current that opposes the line current. In either capacitive or inductive circuits, instantaneous values of power may be very large, but the power actually absorbed is essentially zero, since only resistance dissipates heat (absorbs power). Both inductance and capacitance return the power to the source.

The current in an a-c circuit rises to positive or negative peaks and falls to zero many times per second. Power is stored in the magnetic field of a coil or in the charge of a capacitor. In either case, the power is returned to the source when the current changes direction.

Apparent power is the product of line voltage and line current in a circuit, or $P_{ap} = E_sI_t$. It is called apparent power because the reactive components (inductors and capacitors) only appear to consume power. The power is actually returned to the source. (Bear in mind, of course, that some resistance is associated with inductors and capacitors, and resistance does absorb power.)

Any component that has resistance, such as a resistor or the wiring of an inductor, consumes power. Such power is not returned to the source, because it is dissipated as heat. Power consumed in the circuit is called true power, or average power. The two terms are interchangeable, but we shall use the term average power, since the "overall" value is more meaningful than the instantaneous values of power appearing in the circuit during a complete cycle.

In terms of the dissipation of power as heat in a circuit, define:

(a) Apparent power. _____

_____

(b) Average power. _____

_____

- - - - - - - - - -

(a) Apparent power includes both power that is returned to the source and power that is dissipated as heat.
(b) Average power is power that is dissipated as heat.

79. What kind of power is associated with purely resistive circuits?

_____ With reactive circuits (excluding resistance)?

_____

- - - - - - - - - -

Average power; apparent power

80. Although not all apparent power is consumed by the circuit, it must be considered in design, because the alternator does deliver the power. The average power consumption may be small, but instantaneous values of voltage and current are often very large. Apparent power is an important design consideration, especially in assessing the amount of insulation necessary. Can you tell why? _____

_____

- - - - - - - - - -

Insulation must be sufficient to withstand instantaneous values of voltage and current, which may be very large.

81. In an a-c circuit that includes both reactance and resistance, some power is consumed by the load and some is returned to the source. How much of each depends on the phase angle, since current normally leads or lags voltage by some angle. From our discussions of L-C-R circuits, we know the ratio R/Z to be the cosine of the phase angle θ. Therefore, it is easy to calculate average power in an L-C-R circuit:

$$P = EI \cos \theta ,$$

where E = effective value of the voltage across the circuit,

   I = effective value of current in the circuit,

   θ = phase angle between voltage and current, and

   P = average power absorbed by the circuit.

The equation for average power in a purely resistive circuit is P = EI.

In an L-C-R circuit, the equation for average power is _____.

- - - - - - - - - -

$P = EI \cos \theta$

82. From the trigonometric tables in Appendix II, what is $\cos \theta$ for a resistive circuit in which voltage and current are in phase? _____

- - - - - - - - - -

1   (You should have looked up the cosine for zero degrees.)

83. In a purely reactive circuit, current and voltage are _____ degrees out of phase.

- - - - - - - - - -

90

84. What is cos θ for a purely reactive circuit? _____

\- \- \- \- \- \- \- \- \- \-

0

85. No power is consumed in a purely reactive circuit (which exists only in theory); it is all returned to the source. Average power is equal to EI cos θ.
   Since cos θ is zero, what is the average power in a purely reactive circuit? _____

\- \- \- \- \- \- \- \- \- \-

zero   (Note: Anything multiplied by zero is zero.)

86. What is cos θ in a purely resistive circuit? _____

\- \- \- \- \- \- \- \- \- \-

1

87. In a resistive circuit, P = EI, because cos θ is 1 and need not be considered. In most cases, the phase angle will be neither 90 nor zero, but somewhere between those extremes. Suppose an L-C-R circuit has a source voltage of 400 v., line current is 2 a., and current leads voltage by 60 degrees.

   What is the average power? _____

\- \- \- \- \- \- \- \- \- \-

400 w.   (400 x 2 x 0.5000)

88. Assume that an L-C-R circuit has a source voltage of 200 v., line current is 3 a., and current lags voltage by 31.8 degrees.

   What is the average power? _____

\- \- \- \- \- \- \- \- \- \-

509.94 w.   (200 x 3 x 0.8499)

89. E = 100 v., I = 2 a., and θ = 58.4.

   What is the average power? _____

\- \- \- \- \- \- \- \- \- \-

104.8 w.   (100 x 2 x 0.5240)

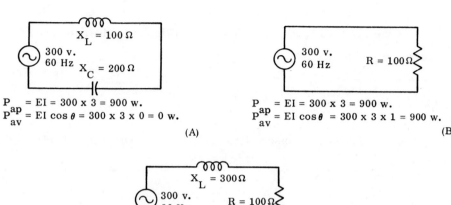

P_{ap} = EI = 300 x 3 = 900 w.
P_{av}^{ap} = EI cos θ = 300 x 3 x 0 = 0 w.

(A)

P_{ap} = EI = 300 x 3 = 900 w.
P_{av}^{ap} = EI cos θ = 300 x 3 x 1 = 900 w.

(B)

P_{ap} = EI = 300 x 1.34 = 401 w.

P_{av} = EI cos θ = 300 x 1.34 x .446 = 179 w.

(C)

Figure 11-10. Power relationships in a-c circuits.

Refer to Figure 11-10 for Frames 90 through 96.

90. In reactive a-c circuits, the relative amounts of apparent power and average power are an important consideration for efficiency and circuit design. Figure 11-10 illustrates some comparisons between apparent and
and average power. The circuit in View A is purely (inductive/capacitive/

reactive) _____.

- - - - - - - - - -

reactive

91. How much power is returned to the source in View A? _____

- - - - - - - - - -

900 w.

92. How much power is returned to the source in View B? _____

- - - - - - - - - -

None.

93. An L-C-R circuit is shown in which View? _____

- - - - - - - - - - -

C

94. What is the circuit impedance in View C? (Hint: Consider the square root of 50,000 to be 223.7.) _____

- - - - - - - - - -

223.7Ω

95. In View C, what is the phase relationship between current and voltage?

_____

- - - - - - - - - -

Current lags voltage by 63.5 degrees.

96. You have probably noticed that the average power in each circuit shown differs from apparent power by the factor of $\cos\theta$. Thus, $\cos\theta$ represents the ratio of apparent power to average power. For example, if the phase angle is 60 degrees, $\cos\theta$ is 0.5000. This ratio is called the power factor. Since 0.5000 also represents 50 percent, we could say that 50 percent of the power in a circuit is actually consumed. This percentage is also a measure of the reactance present in the circuit.
    Write the equation for the power factor (P. F.) when the phase angle is known. _____

- - - - - - - - - -

P. F. = $\cos\theta$

97. What ratio is expressed by $\cos\theta$? _____

- - - - - - - - - -

The ratio (called the power factor) of apparent power to average power.

98. The power factor is always some number within what range? _____

- - - - - - - - - -

0 to 1

99. The apparent power of a circuit is 600 w. and the average power is 450 w.

The power factor is _____, so _____ percent of the power delivered to the circuit is actually dissipated.

- - - - - - - -

0.75; 75

100. Current leads voltage by 35.9 degrees in a given circuit. The apparent power is 1,000 w. The power actually dissipated in the circuit is

_____.

- - - - - - - - - -

810 w.

In this chapter you have learned to calculate voltage, current, and impedance in a series L-C-R circuit. You have also learned to calculate those values in parallel L-C circuits and parallel L-C-R circuits.

You have seen that the calculation of power in a-c circuits requires the application of a power factor, since reactance does not consume power. Thus, you have had to deal with the concepts of average power and apparent power.

When you feel that you understand the material in this chapter, turn to the Self-Test.

Self-Test

The following questions will test your understanding of Chapter Eleven. Write your answers on a separate sheet of paper and check them with the answers provided following the test.

1. How does Kirchhoff's Law of Voltages apply to L-C-R circuits?

2. What two things must be known before vectors can be used to add voltages or currents in an L-C-R circuit?

3. The source voltage of an L-C-R circuit is 400 v. and the line current is 25 a. What is the circuit impedance?

4. In an L-C-R circuit, $X_L = 20\Omega$, $X_C = 90\Omega$, and $R = 70\Omega$. What is $\tan\theta$?

5. In the circuit described in Question 4, what is the phase relationship, including the phase angle, between current and voltage?

6. In an L-C-R circuit, $R = 1600\,\Omega$, $X_L = 2400\Omega$, $X_C = 1200\Omega$, and $Z = 2000\Omega$. What is $\cos\theta$?

7. In an L-C-R circuit, what is the phase relationship between current and voltage if X_L is greater than X_C?

8. Under what conditions is an L-C-R circuit at resonance?

9. In a series L-C-R circuit at resonance, impedance is _____ and line current is _____.

10. In a parallel L-C circuit, what happens to line current on either side of resonance?

11. In a parallel L-C-R circuit at resonance, the impedance is effectively equal to what?

12. What kind of power is delivered to an a-c circuit by the alternator?

13. What kind of power is dissipated as heat?

14. Define the power factor as a ratio.

15.

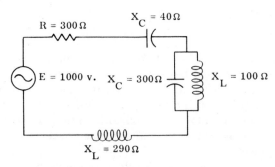

Refer to the schematic diagram and answer the following questions.

    (a) What is Z_t?
    (b) What is I_t?
    (c) What is the phase relationship, including the phase angle, between the line current and the source voltage?
    (d) What is the current through the 100-ohm inductor?
    (e) What is the apparent power of the circuit?
    (f) What is the average power of the circuit?

Answers

If your answers to the test questions do not agree with those given below, review the frames indicated in parentheses. Answers to some questions are reflected in various parts of the chapter, so equations rather than specific frame numbers are given.

1. The sum of instantaneous voltage drops around a closed circuit is equal to the total applied voltage.　(5-7)

2. Effective values and phase angles.　(6)

3. $16\,\Omega$　(10)

4. 1.0　(12)

5. Current leads voltage by 31.0 degrees.　(13, 25)

6. 0.8000　(15)

7. Current lags voltage.　(38)

8. When $X_L = X_C$.　(48)

9. minimum; maximum　(49)

10. Line current increases.　(61, 63)

11. Resistance　(77)

12. Apparent power.   (78–80)

13. Average power.   (78)

14. The power factor is the ratio of average power to apparent power.   (96)

15. (a) $Z_t = 500\Omega$

   To find the impedance of the parallel portion of the circuit:

$$Z = \frac{X_L \times (-X_C)}{X_L - X_C}$$

$$= \frac{100 \times (-300)}{100 - 300} = \frac{-30000}{-200}$$

$$= 150\Omega$$

$$Z_t = \sqrt{R^2 + (X_L - X_C)^2}$$

$$= \sqrt{300^2 + 400^2} = \sqrt{90000 + 160000}$$

$$= \sqrt{250000} = 500\Omega$$

(b) 2 a. $(I_t = \dfrac{E}{Z_t})$

(c) The line current lags the source voltage by 53.1 degrees.   (The net reactance is inductive.   Tan θ = 1.3333.)

(d) 3 a.   (The voltage across the parallel branch is 300 v.   I = $\dfrac{E}{X_L}$ = 3 a.)

(e) 2000 w. $(P_{ap} = EI)$

(f) 1200 w.   (P_{av} = EI cos θ.   If you calculated instead the power dissipated by the resistor (I^2R), good thinking.)

APPENDIX I
Simple Equations

Problems in electricity are solved by the manipulation of mathematical equations, or "formulas," in which some things are known and some unknown quantity must be determined. For example, if current and resistance are known, voltage can be found by setting up and then solving the appropriate equation.

If you have a good head for figures, you could eventually solve the problems in this book by arithmetic, but the process would be laborious and time-consuming. Algebra, or at least that part of algebra dealing with simple equations, is a faster and easier way to solve those problems.

When you have finished this appendix, you will know how to:

- solve for an unknown quantity when any number of known quantities are given; and

- manipulate the terms of an equation to produce other useful equations.

If you wish to test your knowledge of simple equations before working through this Appendix, go to the Self-Test at the end of this Appendix. If you can correctly solve all the problems given there, go on to Chapter One.

1. The signs of numbers are important in algebra; in fact, you can't solve even simple equations without a good understanding of negative and positive values. Fortunately this subject is not new to you. For example, temperatures may be positive (plus, or +) or negative (minus, or −). The temperature "90 degrees above zero Fahrenheit" is written +90° F. (If a number has no sign, it is understood to be positive.) Using the same form, write the temperature "20 degrees below zero Fahrenheit."

- - - - - - - - -

−20° F

2. In algebra, signed numbers may be added, subtracted, multiplied, or divided; and there are special rules for performing these operations. In arithmetic, a common form for indicating subtraction is

$$
\begin{array}{r}
20 \\
-12 \\
\hline
8
\end{array}
$$

In algebra, this operation is not subtraction but <u>addition</u> of 20 and −12. Any number of positive and negative quantities may be added. Simply combine all numbers with like signs (positive or negative), subtract the smaller from the larger combination, giving the sign of the larger to the result.

$$(+20) + (-12) = 8$$
$$20 - 12 = 8$$

Parentheses were used to show that the two numbers, 20 and −12, were to be added. In algebra this is unnecessary, because any series of positive and negative numbers are assumed to be added unless subtraction is specifically indicated. Here is such a series of numbers.

$$-4 +12 +6 -18$$

Collect the positive and negative numbers.

$$-22 +18$$

What is the algebraic sum of these numbers?

$$+5 +16 -8 -12 +6 \ = \ \underline{\hspace{2cm}}$$

- - - - - - - - - -

+7 (or simply 7)

3. What is the algebraic sum of these numbers?

$$7 +13 -28 +6 -4 \ = \ \underline{\hspace{2cm}}$$

- - - - - - - - - -

−6

4. Letters or other symbols as well as pure numbers are used in algebraic equations. Each letter may have some number preceding it, such as 3b, or −2x. These terms must be combined separately.

$$3b -b +4 -7 +4b$$

In this case, first collect all <u>like terms</u>; that is, those terms that include b and those terms that are pure numbers. Then proceed with the rule for addition.

$$(3b -b +4b) + (+4 -7)$$

$$6b -3$$

The parentheses were used to show the grouping of like terms. They are unnecessary in practice. Add these terms:

$$x +12 -4x +8x -14 \ = \ \underline{\hspace{2cm}}$$

- - - - - - - - - -

5x −2

5. Add these terms: 2y −8y +4 −y +16 = _____

- - - - - - - - - -

−7y +20

6. To subtract in algebra, merely change the sign of the subtrahend (number
   to be subtracted) and proceed as in addition. Subtract 4 from 12:
   12 −4 = 8. Subtract 8a from 15a: 15a −8a = 7a. Subtract 10x from 6x:
   6x −10x = −4x. Subtract 8y from 20y: _____

- - - - - - - - - -

20y −8y = 12y

7. Subtract 9p from 6p: _____

- - - - - - - - - -

−3p

8. If unlike terms are to be subtracted, the result will be a combination of
   terms:

     Subtract 4b from 12: 12 −4b

   Subtract 4z from 628: _____

- - - - - - - - - -

628 −4z

9. Subtract 12 from 16m: _____

- - - - - - - - - -

16m −12

10. You are probably used to the sign "x" to represent multiplication. But if
    you want to multiply x by 2, using the arithmetic sign "x," you have the
    expression x x 2. That's a bit confusing, isn't it? Multiplication is indi-
    cated in a number of ways in algebra. One way is to use a dot (·): 2· x;
    4· 3. Write the expression "8 multiplied by 6." _____

- - - - - - - - - -

8· 6

11. Multiplication is also represented by simply writing the numbers together in the case of mixed numbers. You would not write 4·b; you would simply write 4b. Could you write 44 to mean "four multiplied by four"?

_____

- - - - - - - - - -

No, because 44 is a different number.

12. Parentheses are also used to indicate multiplication.

$$(4 + 12)\ (2x - x) = 16$$

Note that addition was performed inside the parentheses first.

$$(8 + 4)\ (4x - 2x) = \underline{\hspace{2cm}}$$

- - - - - - - - - -

24x

13. $(3x - 2x)(4 - 2) = \underline{\hspace{2cm}}$

- - - - - - - - - - -

2x

14. Multiplication in algebra is the same as in arithmetic, except that you must decide what <u>sign</u> to apply to the product. The product is plus if the terms have like signs and minus if the signs are different.

$$2x \cdot 4 = 8x \qquad\qquad -8x \cdot 3 = -24x$$
$$-4y \cdot 3 = -12y \qquad\qquad -3y \cdot -5 = 15y$$

Multiply these terms: (a) $4x \cdot -2 = \underline{\hspace{1.5cm}}$

(b) $-8p \cdot -4 = \underline{\hspace{1.5cm}}$

(c) $6t \cdot -3 = \underline{\hspace{1.5cm}}$

- - - - - - - - - -

(a) −8x; (b) 32p; (c) −18t

15. The division sign "÷" is not used in algebra, but division can, of course, be indicated. The fraction $\frac{2}{3}$ really means 2 ÷ 3. Division is indicated in algebra by placing the terms as numerators or denominators of fractions. Write the expression "y divided by 4." $\underline{\hspace{3cm}}$

- - - - - - - - - -

$$\frac{y}{4}$$

16. Write as an algebraic equation this statement: "Two times x is equal to four times the quantity ten divided by two." _____

- - - - - - - - -

$$2x = 4 \cdot \frac{10}{2}$$

17. In the expression $\frac{4x}{2}$, 4x is the dividend and 2 is the divisor. If the dividend and divisor have the same sign, the quotient (answer) is positive (+); if the divisor and dividend have opposite signs, the quotient is negative (−).

$$\frac{4}{2} = 2 \qquad \frac{-8}{4} = -2 \qquad \frac{-6x}{-3} = 2x$$

(a) $\frac{-16x}{4} =$ _____ (b) $\frac{25x}{5} =$ _____ (c) $\frac{-12x}{-3} =$ _____

- - - - - - - - - -

(a) −4x; (b) 5x; (c) 4x

18. All the divisors in Frame 17 were pure numbers. Division of literal (b, x) or mixed (3b, 4x) numbers is just as easy. Any number divided by itself is 1. $\frac{x}{x} =$ _____

- - - - - - - - - -

1

19. $\frac{16x}{2x} =$ _____

- - - - - - - - -

8

20. $\frac{24y}{-3y} =$ _____

- - - - - - - - -

−8

21. You probably observed that in the examples above, the literal numbers (x, y) in the numerators and denominators canceled out. To see clearly what actually happened, we must talk about <u>exponents</u>. An exponent indicates the number of times a base is multiplied times itself.

$$4^2 = 4 \cdot 4; \quad 10^3 = 10 \cdot 10 \cdot 10; \quad x^4 = x \cdot x \cdot x \cdot x$$

For example, in the expression $3x^2$, the exponent 2 applies only to the x, not to the 3, so $3x^2 = 3 \cdot x \cdot x$. If x in this example stands for 2, the term $3x^2 = 3 \cdot 2 \cdot 2$, or 12. If we let x represent 3, the term $4x^2 = $ _____ .

- - - - - - - - - -

36    $(4 \cdot x \cdot x = 4 \cdot 3 \cdot 3 = 36)$

22. $2^2 x^3 = $ _____

- - - - - - - - - - -

$4x^3$

23. The first power, exponent 1, is not written because any number raised to the first power is the number itself. $24^1 = $ _____

- - - - - - - - - -

24

24. Now let's look again at those expressions in which the letter canceled out.
$\frac{16x}{2x} = 8; \quad \frac{24y}{-3y} = -8$. Any number divided by itself is equal to one, and that is the reason the letters were eliminated. Mathematically, division is accomplished by the <u>subtraction of exponents.</u> In the first example, $\frac{16x}{2x}$, 16 was divided by 2 in the normal way, but the x's were handled like this: $x^{1-1} = x^0$. The numbers could have been handled in the same manner. Another way of writing 16 is 2^4, and 2 could have been written 2^1. You know that $16 \div 2 = 8$. This expression is also true: $2^{4-1} = 2^3$. Let's divide 16 by 16, using powers of 2.

$$\frac{2^4}{2^4} = 2^{4-4} = 2^0 = \underline{\qquad}$$

- - - - - - - - - -

1

25. Any letter or number raised to the zero power is equal to _____ .

- - - - - - - - - -

1

26. $y^0 = $ _____

- - - - - - - - - -

1

27. $\dfrac{x^4}{x^2} =$ _____

- - - - - - - - - -

x^2

28. Since division is accomplished by subtracting exponents, you might expect that multiplication is accomplished by <u>adding</u> exponents. That's correct. If 2^3 is 8 and 2^4 is 16, you can see that 2^{3+1} is also 16. What is the product of $x^2 \cdot x$? _____

- - - - - - - - - -

x^3

29. (a) $4y \cdot 3y^3 =$ _____   (b) $\dfrac{32x^4}{8x^3} =$ _____

- - - - - - - - - -

(a) $12y^4$;   (b) $4x$

Appendix III presents a short review of the laws of exponents, if you are interested in learning more about them.

30. Algebra begins with simple equations. It continues with not-so-simple equations, but you don't have to go that far to solve the problems in basic electricity. An <u>equation</u>, as the name implies, is a statement of equality. The equation is identified by the equal sign $(=)$, and it means that some term or group of terms on one side of that sign is equal to some term or group of terms on the other side.

$$2 + 3 = 5$$
$$2 + 3 - 5 = 0$$
$$x - 2 = 3$$

Is the following mathematical statement an equation? $2 + 3 = 6$ _____

- - - - - - - - - -

No. (The sum of 3 and 2 is not 6.)

31. You know immediately that the statement $2 + 3 = 6$ is not true. An equation is a true statement. If, in the process of reducing a complex equation to simple form, you end with an expression such as $2 + 3 = 6$, you know that you did something wrong.

Simple equations are used to solve for an unknown quantity, represented by an x or some other literal number or symbol, when other quantities are known, or "given." A 2 can only be a 2 in an equation, because it is

known. A letter may have any value in an equation, but you don't know
what that value is until you solve the equation. Is the following statement

an equation? $3 + x = 7$ _____

- - - - - - - - - -

Yes.

32. The statement $3 + x = 7$ is true, because some value can be assigned to
x that will make it true. In the equation $x + 3 = 8$, what is the value of x?

_____

- - - - - - - - - -

5

33. You know that $x = 5$, because that is the only number that can be added to
3 for a sum of 8. However, you didn't necessarily use algebra to solve
for x. You might have made a series of rapid mental calculations, using
trial and error, before you decided that x was equal to 5. Algebra is a
system by which you can manipulate the terms of an equation in an orderly
manner to arrive at a value for an unknown. The first step in solving an
equation is to <u>collect like terms</u>. For example, suppose that we want to
add the quantities 2a, 10, 4a, and 2. The numbers 10 and 2 are like
terms, since both are real numbers with known values. The terms 2a
and 4a are also like terms, because each is a value of a. Therefore,
$$2a + 4a + 10 + 2 = 6a + 12.$$
We have an equation, because the sum $6a + 12$ is equal to the sum of all
four terms expressed separately. The terms 2x and 4a are <u>not alike</u> because they contain different letters.
$$2a + 4a + 10 + 2 + 2x = 6a + 2x + 12$$
The terms can be expressed in any order, but convention usually places
the purely numerical term last. Add these quantities:
$$3x + 4 + 2y + 4x + 6 = \text{_____}$$

- - - - - - - - - -

$7x + 2y + 10$

34. In solving the problems in this book, you will need to deal with only one
unknown quantity. The equation $x + 3 = 8$ contains such an unknown: x.
The key to solving a simple equation is to get the unknown on one side and
all the known terms on the other side of the equal sign. There is a basic
principle in algebra that allows us to do so. Whatever is done to one side
of an equation must be done to the other side to maintain the equality. We
can see how this works by examining an equation with no unknowns.
$$4 = 4$$

We can add any quantity to both sides, subtract any quantity, multiply by any quantity, divide by any quantity—in short, we can perform any mathematical operation we wish.

$$4 = 4$$

Add 2:	$6 = 6$
Subtract 2:	$2 = 2$
Multiply by 2:	$8 = 8$
Divide by 2:	$2 = 2$
Take the square root:	$2 = 2$
Square each side:	$16 = 16$

Now let's look at an equation that contains an unknown.

$$x + 4 = 12$$

We can isolate x on the left side of the equation by subtracting 4 from both sides of the equation. Subtracting 4 has the same result as adding −4. Since we're working with algebra, it is simpler to think of adding a negative number, since you must change the sign anyway if you subtract.

$$x + 4 - 4 = 12 - 4$$

$$x = 8$$

Let's work the equation again, mentally adding −4 to the left side and writing it on the right side. (We're using this odd method to make a point.)

$$x + 4 = 12$$

$$x = 12 - 4$$

We have in effect moved the 4 from the left side to the right. The two equations are identical, except that the first included +4 on the left, and the second had −4 on the right. Thus, we have discovered another principle of simple equations. Any term may be moved across the equal sign by changing its _____.

- - - - - - - - -

sign

35. In the equation below, move the 3 across the equal sign and write the new equation. $x + 3 = 12$ _____

- - - - - - - - -

$x = 12 - 3$

36. Move the −6 across the equal sign. $x - 6 = 18$ _____

- - - - - - - - -

x = 18 + 6

37. Solve for y: y + 16 = 20. _____

- - - - - - - - - -

y = 4

38. Solve for p: p − 3 = 9. _____

- - - - - - - - - -

p = 12

39. Of course, more than one term containing the unknown might appear in an equation.

$$4p - 2 = 2p + 6$$

In this case, you must get the p's together on one side of the equation and the numbers together on the other side. You can do this mentally, but it's safer to write out each step until you are very familiar with the operations.

$$4p - 2 = 2p + 6$$
$$4p - 2 - 2p = 2p + 6 - 2p$$
$$2p - 2 = 6$$
$$2p - 2 + 2 = 6 + 2$$
$$2p = 8$$

Only one more step remains. We must isolate p (not 2p) on the left.

What mathematical operation will accomplish this? _____

- - - - - - - - - -

Divide each side by 2.

40. 2p = 8. p = _____

- - - - - - - - - -

4

41. 3x − 2 = 2x + 1. x = _____

- - - - - - - - - -

3

42. $6t - 24 = 6 - 4t.$ $t =$ _____

- - - - - - - - - -

$6t + 4t = 6 + 24$
  $10t = 30$
    $t = 3$

43. You do not necessarily have to end your solution with a numerical value for the unknown. You might merely solve for one unknown in terms of others.

$$2z + x = z + y$$

$$2z - z = y - x$$

$$z = y - x$$

The basic equations in electricity start as <u>formulas</u>, which are merely equations in which letters or symbols stand for given values. For example, an essential formula in electricity is $E = IR$, in which E represents voltage in volts, I is current in amperes, and R is resistance in ohms. If any two values are known, the third can be found, but it may be necessary to develop a variation of the formula. If E and R are known, I may be found after the proper formula is arranged.

$E = IR$ Divide through by R:

$\dfrac{E}{R} = \dfrac{IR}{R}$ The R's on the right cancel out.

$\dfrac{E}{R} = I$

Develop the formula for R. _____

- - - - - - - - - -

$\dfrac{E}{I} = R$ (Note: By convention, since people are used to reading from left to right, the unknown to be solved is placed on the left side of the equal sign: $R = \dfrac{E}{I}$.)

44. Parentheses are often used in algebra to indicate multiplication. <u>Every</u> term inside the parentheses must be multiplied by the term preceding the parentheses.

$$3(x + 2) = 3x + 6 \qquad 4(x - 3) = \underline{\qquad\qquad}$$

- - - - - - - - - -

$4x - 12$

45. An important rule in algebra is: Always do the work inside the parenthe-
    ses first.   This operation removes the parentheses, and you can then
    proceed.

    $$8 + 2(x - 1)$$

    The expression $2(x - 1)$ is a single term, but if you are careless, you
    might add 8 and 2 first, to arrive at a wrong expression, $10(x - 1)$.
    What is the correct expression that results from removing the paren-

    theses? _____

    - - - - - - - - - -

    $8 + 2x - 2$, or $6 + 2x$

46. Always do multiplications, as indicated by parentheses or by the symbol
    $\cdot$, before you add or subtract.   Consider this example.

    $$3 \cdot 8 - 4 = 20$$

    If you subtracted 4 from 8 (or algebraically added 8 and -4) first, your
    answer would be 12, which is incorrect. You must multiply 3 and 8 first,
    and then subtract 4, which gives the correct answer of 20.
        There are five terms in the left-hand side of the equation below.   Cir-
    cle the term or terms you should collect or simplify <u>first</u>.

    $$5x \quad - \quad 3x \quad + \quad 22 \quad - \quad 12 \quad + \quad 6(3x - 2) \quad = \quad 78$$

    - - - - - - - - - -

    $+6(3x - 2)$.   Perform multiplication before addition or subtraction.   The
    simplified equation is $5x - 3x + 22 - 12 + 18x - 12 = 78$, or $20x - 2 = 78$.
    $(x = 4)$

47. You have probably discovered a simple way to check the correctness of
    your answer in the problems you have worked so far. You simply substi-
    tute the value of the unknown and see if you have a true equation.

    $$3x + 4 = 16$$
    $$x = 4$$
    $$3 \cdot 4 + 4 = 16$$
    $$12 + 4 = 16$$
    $$16 = 16$$

    If you end with the same number on either side of the equal sign, your
    solution is correct.   Is the solution of this equation correct?

    $$4y + 3 = y + 12; \quad y = 5$$

    _____

    - - - - - - - - - -

    No, because substitution of 5 in the equation will result in $23 = 17$.   The
    true value of y is 3: $15 = 15$.

48. Any operation may be performed on one side of an equation if the same operation is performed on the other side. In the following equation, the exponents can be removed by taking the <u>square root</u> of each side.

$$x^2 = 16 + 9$$
$$x^2 = 25$$
$$x = 5$$

Many equations in electricity require that the square root be extracted. What must be done to the equation $Z^2 = R^2 + X^2$ to isolate Z on the left?

_ _ _ _ _ _ _ _ _ _

Take the square root of each side.

49. $R = 3$ and $X = 4$. Solve for Z using the formula $Z^2 = R^2 + X^2$.

_ _ _ _ _ _ _ _ _

$$\sqrt{Z^2} = \sqrt{R^2 + X^2}$$
$$Z = \sqrt{3^2 + 4^2} = \sqrt{9 + 16} = \sqrt{25} = 5$$

50. One formula for power that is frequently used in electricity is $P = I^2R$. You can develop the formula for finding current (I) when power (P) and resistance (R) are known simply by solving the basic equation for I.

$P = I^2R$ Divide by R to isolate I.

$\dfrac{P}{R} = \dfrac{I^2R}{R}$ The R's on the right cancel out.

$\dfrac{P}{R} = I^2$ Take the square root of both sides.

$\sqrt{\dfrac{P}{R}} = \sqrt{I^2}$, or $I = \sqrt{\dfrac{P}{R}}$

Another formula for power is $P = \dfrac{E^2}{R}$. The first step is shown, because it is somewhat difficult, and there is a short-cut that you will learn later.

$$E^2 = PR; \quad E = \underline{\hspace{3cm}}$$

_ _ _ _ _ _ _ _ _

$E = \sqrt{PR}$

51. A useful short-cut in algebra is cross-multiplication, which can be used any time there is a simple fraction on either side of the equal sign. (Remember that any single term is also the fraction produced by including the denominator 1.)

$$P = \frac{E^2}{R} , \text{ or } \frac{P}{1} = \frac{E^2}{R}$$

In cross-multiplication, you multiply each numerator (upper term of the fraction) by the opposite denominator (lower term). Since any term multiplied by 1 is the same term, the 1 need not be written.

$$E^2 = PR$$

Solve for x in the equation below.

$$\frac{x}{4} = \frac{y}{2}$$

- - - - - - - - -

$2x = 4y$

$$\frac{2x}{2} = \frac{4y}{2}$$

$x = 2y$

52. One formula for current is $I = \frac{E}{R}$. What is the formula for voltage (E)?

- - - - - - - - -

$E = IR$

Self-Test

The following problems will test your understanding of Appendix I. Answers are provided following the test.

1. Solve for a: $\frac{6a^2}{2a} = 9$

2. Solve for x: $2x + 3 - 6 = 9 - x$

3. Solve for R: $P = I^2R$

4. Solve for p: $2p + 3(p + 2) = 21$

5. Solve for c: $c^2 = a^2 + b^2$

6. Solve for z: $z^2 = R^2 + x^2$

7. Solve for x: $2x = \frac{3y}{2}$

8. Solve for X_L: $L = \dfrac{X_L}{2\pi f}$

9. Solve for E: $I = \dfrac{E}{R}$

10. Solve for L: $t = \dfrac{L}{R}$

Answers

If your answers do not agree with those given, review the frames indicated in parentheses after each answer. If you solved all problems correctly, you are ready to solve any equation in this book.

1. $a = 3$ (27, 41)

2. $x = 4$ (41)

3. $R = \dfrac{P}{I^2}$ (43)

4. $p = 3$ (46)

5. $c = \sqrt{a^2 + b^2}$ (48–50)

6. $z = \sqrt{R^2 + X^2}$ (48–50)

7. $x = \dfrac{3y}{4}$ (51)

8. $X_L = L2\pi f$ (or $2\pi fL$)   (51, 52)

9. $E = IR$ (52)

10. $L = tR$ (52)

APPENDIX II
Trigonometric Functions

Deg.	Sin	Tan	Cot	Cos	Deg.	Deg.	Sin	Tan	Cot	Cos	Deg.
0.0	0.00000	0.00000	∞	1.0000	90.0	.5	.07846	.07870	12.706	.9969	.5
.1	.00175	.00175	573.0	1.0000	.9	.6	.08020	.08046	12.429	.9968	.4
.2	.00349	.00349	286.5	1.0000	.8	.7	.08194	.08221	12.163	.9966	.3
.3	.00524	.00524	191.0	1.0000	.7	.8	.08368	.08397	11.909	.9965	.2
.4	.00698	.00698	143.24	1.0000	.6	.9	.08542	.08573	11.664	.9963	.1
.5	.00873	.00873	114.59	1.0000	.5	5.0	0.08716	0.08749	11.430	0.9962	85.0
.6	.01047	.01047	95.49	0.9999	.4	.1	.08889	.08925	11.205	.9960	.9
.7	.01222	.01222	81.85	.9999	.3	.2	.09063	.09101	10.988	.9959	.8
.8	.01396	.01396	71.62	.9999	.2	.3	.09237	.09277	10.780	.9957	.7
.9	.01571	.01571	63.66	.9999	.1	.4	.09411	.09453	10.579	.9956	.6
1.0	0.01745	0.01746	57.29	0.9998	89.0	.5	.09585	.09629	10.385	.9954	.5
.1	.01920	.01920	52.08	.9998	.9	.6	.09758	.09805	10.199	.9952	.4
.2	.02094	.02095	47.74	.9998	.8	.7	.09932	.09981	10.019	.9951	.3
.3	.02269	.02269	44.07	.9997	.7	.8	.10106	.10158	9.845	.9949	.2
.4	.02443	.02444	40.92	.9997	.6	.9	.10279	.10334	9.677	.9947	.1
.5	.02618	.02619	38.19	.9997	.5	6.0	0.10453	0.10510	9.514	0.9945	84.0
.6	.02792	.02793	35.80	.9996	.4	.1	.10626	.10687	9.357	.9943	.9
.7	.02967	.02968	33.69	.9996	.3	.2	.10800	.10863	9.205	.9942	.8
.8	.03141	.03143	31.82	.9995	.2	.3	.10973	.11040	9.058	.9940	.7
.9	.03316	.03317	30.14	.9995	.1	.4	.11147	.11217	8.915	.9938	.6
2.0	0.03490	0.03492	28.64	0.9994	88.0	.5	.11320	.11394	8.777	.9936	.5
.1	.03664	.03667	27.27	.9993	.9	.6	.11494	.11570	8.643	.9934	.4
.2	.03839	.03842	26.03	.9993	.8	.7	.11667	.11747	8.513	.9932	.3
.3	.04013	.04016	24.90	.9992	.7	.8	.11840	.11924	8.386	.9930	.2
.4	.04188	.04191	23.86	.9991	.6	.9	.12014	.12101	8.264	.9928	.1
.5	.04362	.04366	22.90	.9990	.5	7.0	0.12187	0.12278	8.144	0.9925	83.0
.6	.04536	.04541	22.02	.9990	.4	.1	.12360	.12456	8.028	.9923	.9
.7	.04711	.04716	21.20	.9989	.3	.2	.12533	.12633	7.916	.9921	.8
.8	.04885	.04891	20.45	.9988	.2	.3	.12706	.12810	7.806	.9919	.7
.9	.05059	.05066	19.74	.9987	.1	.4	.12880	.12988	7.700	.9917	.6
3.0	0.05234	0.05241	19.081	0.9986	87.0	.5	.13053	.13165	7.596	.9914	.5
.1	.05408	.05416	18.464	.9985	.9	.6	.13226	.13343	7.495	.9912	.4
.2	.05582	.05591	17.886	.9984	.8	.7	.13399	.13521	7.396	.9910	.3
.3	.05756	.05766	17.343	.9983	.7	.8	.13572	.13698	7.300	.9907	.2
.4	.05931	.05941	16.832	.9982	.6	.9	.13744	.13876	7.207	.9905	.1
.5	.06105	.06116	16.350	.9981	.5	8.0	0.13917	0.14054	7.115	0.9903	82.0
.6	.06279	.06291	15.895	.9980	.4	.1	.14090	.14232	7.026	.9900	.9
.7	.06453	.06467	15.464	.9979	.3	.2	.14263	.14410	6.940	.9898	.8
.8	.06627	.06642	15.056	.9978	.2	.3	.14436	.14588	6.855	.9895	.7
.9	.06802	.06817	14.669	.9977	.1	.4	.14608	.14767	6.772	.9893	.6
4.0	0.06976	0.06993	14.301	0.9976	86.0	.5	.14781	.14945	6.691	.9890	.5
.1	.07150	.07168	13.951	.9974	.9	.6	.14954	.15124	6.612	.9888	.4
.2	.07324	.07344	13.617	.9973	.8	.7	.15126	.15302	6.535	.9885	.3
.3	.07498	.07519	13.300	.9972	.7	.8	.15299	.15481	6.460	.9882	.2
.4	.07672	.07695	12.996	.9971	85.6	.9	.15471	.15660	6.386	.9880	81.1
Deg.	Cos	Cot	Tan	Sin	Deg.	Deg.	Cos	Cot	Tan	Sin	Deg.

Deg.	Sin	Tan	Cot	Cos	Deg.	Deg.	Sin	Tan	Cot	Cos	Deg.
9.0	0.15643	0.15838	6.314	0.9877	81.0	.5	.2334	.2401	4.165	.9724	.5
.1	.15816	.16017	6.243	.9874	.9	.6	.2351	.2419	4.134	.9720	.4
.2	.15988	.16196	6.174	.9871	.8	.7	.2368	.2438	4.102	.9715	.3
.3	.16160	.16376	6.107	.9869	.7	.8	.2385	.2456	4.071	.9711	.2
.4	.16333	.16555	6.041	.9866	.6	.9	.2402	.2475	4.041	.9707	.1
.5	.16505	.16734	5.976	.9863	.5	14.0	0.2419	0.2493	4.011	0.9703	76.0
.6	.16677	.16914	5.912	.9860	.4	.1	.2436	.2512	3.981	.9699	.9
.7	.16849	.17093	5.850	.9857	.3	.2	.2453	.2530	3.952	.9694	.8
.8	.17021	.17273	5.789	.9854	.2	.3	.2470	.2549	3.923	.9690	.7
.9	.17193	.17453	5.730	.9851	.1	.4	.2487	.2568	3.895	.9686	.6
10.0	0.1736	0.1763	5.671	0.9848	80.0	.5	.2504	.2586	3.867	.9681	.5
.1	.1754	.1781	5.614	.9845	.9	.6	.2521	.2605	3.839	.9677	.4
.2	.1771	.1799	5.558	.9842	.8	.7	.2538	.2623	3.812	.9673	.3
.3	.1788	.1817	5.503	.9839	.7	.8	.2554	.2642	3.785	.9668	.2
.4	.1805	.1835	5.449	.9836	.6	.9	.2571	.2661	3.758	.9664	.1
.5	.1822	.1853	5.396	.9833	.5	15.0	0.2588	0.2679	3.732	0.9659	75.0
.6	.1840	.1871	5.343	.9829	.4	.1	.2605	.2698	3.706	.9655	.9
.7	.1857	.1890	5.292	.9826	.3	.2	.2622	.2717	3.681	.9650	.8
.8	.1874	.1908	5.242	.9823	.2	.3	.2639	.2736	3.655	.9646	.7
.9	.1891	.1926	5.193	.9820	.1	.4	.2656	.2754	3.630	.9641	.6
11.0	0.1908	0.1944	5.145	0.9816	79.0	.5	.2672	.2773	3.606	.9636	.5
.1	.1925	.1962	5.097	.9813	.9	.6	.2689	.2792	3.582	.9632	.4
.2	.1942	.1980	5.050	.9810	.8	.7	.2706	.2811	3.558	.9627	.3
.3	.1959	.1998	5.005	.9806	.7	.8	.2723	.2830	3.534	.9622	.2
.4	.1977	.2016	4.959	.9803	.6	.9	.2740	.2849	3.511	.9617	.1
.5	.1994	.2035	4.915	.9799	.5	16.0	0.2756	0.2867	3.487	0.9613	74.0
.6	.2011	.2053	4.872	.9796	.4	.1	.2773	.2886	3.465	.9608	.9
.7	.2028	.2071	4.829	.9792	.3	.2	.2790	.2905	3.442	.9603	.8
.8	.2045	.2089	4.787	.9789	.2	.3	.2807	.2924	3.420	.9598	.7
.9	.2062	.2107	4.745	.9785	.1	.4	.2823	.2943	3.398	.9593	.6
12.0	0.2079	0.2126	4.705	0.9781	78.0	.5	.2840	.2962	3.376	.9588	.5
.1	.2096	.2144	4.665	.9778	.9	.6	.2857	.2981	3.354	.9583	.4
.2	.2113	.2162	4.625	.9774	.8	.7	.2874	.3000	3.333	.9578	.3
.3	.2130	.2180	4.586	.9770	.7	.8	.2890	.3019	3.312	.9573	.2
.4	.2147	.2199	4.548	.9767	.6	.9	.2907	.3038	3.291	.9568	.1
.5	.2164	.2217	4.511	.9763	.5	17.0	0.2924	0.3057	3.271	0.9563	73.0
.6	.2181	.2235	4.474	.9759	.4	.1	.2940	.3076	3.251	.9558	.9
.7	.2198	.2254	4.437	.9755	.3	.2	.2957	.3096	3.230	.9553	.8
.8	.2215	.2272	4.402	.9751	.2	.3	.2974	.3115	3.211	.9548	.7
.9	.2233	.2290	4.366	.9748	.1	.4	.2990	.3134	3.191	.9542	.6
13.0	0.2250	0.2309	4.331	0.9744	77.0	.5	.3007	.3153	3.172	.9537	.5
.1	.2267	.2327	4.297	.9740	.9	.6	.3024	.3172	3.152	.9532	.4
.2	.2284	.2345	4.264	.9736	.8	.7	.3040	.3191	3.133	.9527	.3
.3	.2300	.2364	4.230	.9732	.7	.8	.3057	.3211	3.115	.9521	.2
.4	.2317	.2382	4.198	.9728	76.6	.9	.3074	.3230	3.096	.9516	72.1
Deg.	Cos	Cot	Tan	Sin	Deg.	Deg.	Cos	Cot	Tan	Sin	Deg.

Deg.	Sin	Tan	Cot	Cos	Deg.	Deg.	Sin	Tan	Cot	Cos	Deg.
18.0	0.3090	0.3249	3.078	0.9511	72.0	.5	.3827	.4142	2.414	.9239	.5
.1	.3107	.3269	3.060	.9505	.9	.6	.3843	.4163	2.402	.9232	.4
.2	.3123	.3288	3.042	.9500	.8	.7	.3859	.4183	2.391	.9225	.3
.3	.3140	.3307	3.024	.9494	.7	.8	.3875	.4204	2.379	.9219	.2
.4	.3156	.3327	3.006	.9489	.6	.9	.3891	.4224	2.367	.9212	.1
.5	.3173	.3346	2.989	.9483	.5	23.0	0.3907	0.4245	2.356	0.9205	67.0
.6	.3190	.3365	2.971	.9478	.4	.1	.3923	.4265	2.344	.9198	.9
.7	.3206	.3385	2.954	.9472	.3	.2	.3939	.4286	2.333	.9191	.8
.8	.3223	.3404	2.937	.9466	.2	.3	.3955	.4307	2.322	.9184	.7
.9	.3239	.3424	2.921	.9461	.1	.4	.3971	.4327	2.311	.9178	.6
19.0	0.3256	0.3443	2.904	0.9455	71.0	.5	.3987	.4348	2.300	.9171	.5
.1	.3272	.3463	2.888	.9449	.9	.6	.4003	.4369	2.289	.9164	.4
.2	.3289	.3482	2.872	.9444	.8	.7	.4019	.4390	2.278	.9157	.3
.3	.3305	.3502	2.856	.9438	.7	.8	.4035	.4411	2.267	.9150	.2
.4	.3322	.3522	2.840	.9432	.6	.9	.4051	.4431	2.257	.9143	.1
.5	.3338	.3541	2.824	.9426	.5	24.0	0.4067	0.4452	2.246	0.9135	66.0
.6	.3355	.3561	2.808	.9421	.4	.1	.4083	.4473	2.236	.9128	.9
.7	.3371	.3581	2.793	.9415	.3	.2	.4099	.4494	2.225	.9121	.8
.8	.3387	.3600	2.778	.9409	.2	.3	.4115	.4515	2.215	.9114	.7
.9	.3404	.3620	2.762	.9403	.1	.4	.4131	.4536	2.204	.9107	.6
20.0	0.3420	0.3640	2.747	0.9397	70.0	.5	.4147	.4557	2.194	.9100	.5
.1	.3437	.3659	2.733	.9391	.9	.6	.4163	.4578	2.184	.9092	.4
.2	.3453	.3679	2.718	.9385	.8	.7	.4179	.4599	2.174	.9085	.3
.3	.3469	.3699	2.703	.9379	.7	.8	.4195	.4621	2.164	.9078	.2
.4	.3486	.3719	2.689	.9373	.6	.9	.4210	.4642	2.154	.9070	.1
.5	.3502	.3739	2.675	.9367	.5	25.0	0.4226	0.4663	2.145	0.9063	65.0
.6	.3518	.3759	2.660	.9361	.4	.1	.4242	.4684	2.135	.9056	.9
.7	.3535	.3779	2.646	.9354	.3	.2	.4258	.4706	2.125	.9048	.8
.8	.3551	.3799	2.633	.9348	.2	.3	.4274	.4727	2.116	.9041	.7
.9	.3567	.3819	2.619	.9342	.1	.4	.4289	.4748	2.106	.9033	.6
21.0	0.3584	0.3839	2.605	0.9336	69.0	.5	.4305	.4770	2.097	.9026	.5
.1	.3600	.3859	2.592	.9330	.9	.6	.4321	.4791	2.087	.9018	.4
.2	.3616	.3879	2.578	.9323	.8	.7	.4337	.4813	2.078	.9011	.3
.3	.3633	.3899	2.565	.9317	.7	.8	.4352	.4834	2.069	.9003	.2
.4	.3649	.3919	2.552	.9311	.6	.9	.4368	.4856	2.059	.8996	.1
.5	.3665	.3939	2.539	.9304	.5	26.0	0.4384	0.4877	2.050	0.8988	64.0
.6	.3681	.3959	2.526	.9298	.4	.1	.4399	.4899	2.041	.8980	.9
.7	.3697	.3979	2.513	.9291	.3	.2	.4415	.4921	2.032	.8973	.8
.8	.3714	.4000	2.500	.9285	.2	.3	.4431	.4942	2.023	.8965	.7
.9	.3730	.4020	2.488	.9278	.1	.4	.4446	.4964	2.014	.8957	.6
22.0	0.3746	0.4040	2.475	0.9272	68.0	.5	.4462	.4986	2.006	.8949	.5
.1	.3762	.4061	2.463	.9265	.9	.6	.4478	.5008	1.997	.8942	.4
.2	.3778	.4081	2.450	.9259	.8	.7	.4493	.5029	1.988	.8934	.3
.3	.3795	.4101	2.438	.9252	.7	.8	.4509	.5051	1.980	.8926	.2
.4	.3811	.4122	2.426	.9245	67.6	.9	.4524	.5073	1.971	.8918	63.1
Deg.	Cos	Cot	Tan	Sin	Deg.	Deg.	Cos	Cot	Tan	Sin	Deg.

Deg.	Sin	Tan	Cot	Cos	Deg.	Deg.	Sin	Tan	Cot	Cos	Deg.
27.0	0.4540	0.5095	1.963	0.8910	63.0	.5	.5225	.6128	1.6319	.8526	.5
.1	.4555	.5117	1.954	.8902	.9	.6	.5240	.6152	1.6255	.8517	.4
.2	.4571	.5139	1.946	.8894	.8	.7	.5255	.6176	1.6191	.8508	.3
.3	.4586	.5161	1.937	.8886	.7	.8	.5270	.6200	1.6128	.8499	.2
.4	.4602	.5184	1.929	.8878	.6	.9	.5284	.6224	1.6066	.8490	.1
.5	.4617	.5206	1.921	.8870	.5	32.0	0.5299	0.6249	1.6003	0.8480	58.0
.6	.4633	.5228	1.913	.8862	.4	.1	.5314	.6273	1.5941	.8471	.9
.7	.4648	.5250	1.905	.8854	.3	.2	.5329	.6297	1.5880	.8462	.8
.8	.4664	.5272	1.897	.8846	.2	.3	.5344	.6322	1.5818	.8453	.7
.9	.4679	.5295	1.889	.8838	.1	.4	.5358	.6346	1.5757	.8443	.6
28.0	0.4695	0.5317	1.881	0.8829	62.0	.5	.5373	.6371	1.5697	.8434	.5
.1	.4710	.5340	1.873	.8821	.9	.6	.5388	.6395	1.5637	.8425	.4
.2	.4726	.5362	1.865	.8813	.8	.7	.5402	.6420	1.5577	.8415	.3
.3	.4741	.5384	1.857	.8805	.7	.8	.5417	.6445	1.5517	.8406	.2
.4	.4756	.5407	1.849	.8796	.6	.9	.5432	.6469	1.5458	.8396	.1
.5	.4772	.5430	1.842	.8788	.5	33.0	0.5446	0.6494	1.5399	0.8387	57.0
.6	.4787	.5452	1.834	.8780	.4	.1	.5461	.6519	1.5340	.8377	.9
.7	.4802	.5475	1.827	.8771	.3	.2	.5476	.6544	1.5282	.8368	.8
.8	.4818	.5498	1.819	.8763	.2	.3	.5490	.6569	1.5224	.8358	.7
.9	.4833	.5520	1.811	.8755	.1	.4	.5505	.6594	1.5166	.8348	.6
29.0	0.4848	0.5543	1.804	0.8746	61.0	.5	.5519	.6619	1.5108	.8339	.5
.1	.4863	.5566	1.797	.8738	.9	.6	.5534	.6644	1.5051	.8329	.4
.2	.4879	.5589	1.789	.8729	.8	.7	.5548	.6669	1.4994	.8320	.3
.3	.4894	.5612	1.782	.8721	.7	.8	.5563	.6694	1.4938	.8310	.2
.4	.4909	.5635	1.775	.8712	.6	.9	.5577	.6720	1.4882	.8300	.1
.5	.4924	.5658	1.767	.8704	.5	34.0	0.5592	0.6745	1.4826	0.8290	56.0
.6	.4939	.5681	1.760	.8695	.4	.1	.5606	.6771	1.4770	.8281	.9
.7	.4955	.5704	1.753	.8686	.3	.2	.5621	.6796	1.4715	.8271	.8
.8	.4970	.5727	1.746	.8678	.2	.3	.5635	.6822	1.4659	.8261	.7
.9	.4985	.5750	1.739	.8669	.1	.4	.5650	.6847	1.4605	.8251	.6
30.0	0.5000	0.5774	1.7321	0.8660	60.0	.5	.5664	.6873	1.4550	.8241	.5
.1	.5015	.5797	1.7251	.8652	.9	.6	.5678	.6899	1.4496	.8231	.4
.2	.5030	.5820	1.7182	.8643	.8	.7	.5693	.6924	1.4442	.8221	.3
.3	.5045	.5844	1.7113	.8634	.7	.8	.5707	.6950	1.4388	.8211	.2
.4	.5060	.5867	1.7045	.8625	.6	.9	.5721	.6976	1.4335	.8202	.1
.5	.5075	.5890	1.6977	.8616	.5	35.0	0.5736	0.7002	1.4281	0.8192	55.0
.6	.5090	.5914	1.6909	.8607	.4	.1	.5750	.7028	1.4229	.8181	.9
.7	.5105	.5938	1.6842	.8599	.3	.2	.5764	.7054	1.4176	.8171	.8
.8	.5120	.5961	1.6775	.8590	.2	.3	.5779	.7080	1.4124	.8161	.7
.9	.5135	.5985	1.6709	.8581	.1	.4	.5793	.7107	1.4071	.8151	.6
31.0	0.5150	0.6009	1.6643	0.8572	59.0	.5	.5807	.7133	1.4019	.8141	.5
.1	.5165	.6032	1.6577	.8563	.9	.6	.5821	.7159	1.3968	.8131	.4
.2	.5180	.6056	1.6512	.8554	.8	.7	.5835	.7186	1.3916	.8121	.3
.3	.5195	.6080	1.6447	.8545	.7	.8	.5850	.7212	1.3865	.8111	.2
.4	.5210	.6104	1.6383	.8536	58.6	.9	.5864	.7239	1.3814	.8100	54.1
Deg.	Cos	Cot	Tan	Sin	Deg.	Deg.	Cos	Cot	Tan	Sin	Deg.

Deg.	Sin	Tan	Cot	Cos	Deg.	Deg.	Sin	Tan	Cot	Cos	Deg.
36.0	0.5878	0.7265	1.3764	0.8090	54.0	.5	.6494	.8541	1.1708	.7604	.5
.1	.5892	.7292	1.3713	.8080	.9	.6	.6508	.8571	1.1667	.7593	.4
.2	.5906	.7319	1.3663	.8070	.8	.7	.6521	.8601	1.1626	.7581	.3
.3	.5920	.7346	1.3613	.8059	.7	.8	.6534	.8632	1.1585	.7570	.2
.4	.5934	.7373	1.3564	.8049	.6	.9	.6547	.8662	1.1544	.7559	.1
.5	.5948	.7400	1.3514	.8039	.5	41.0	0.6561	0.8693	1.1504	0.7547	49.0
.6	.5962	.7427	1.3465	.8028	.4	.1	.6574	.8724	1.1463	.7536	.9
.7	.5976	.7454	1.3416	.8018	.3	.2	.6587	.8754	1.1423	.7524	.8
.8	.5990	.7481	1.3367	.8007	.2	.3	.6600	.8785	1.1383	.7513	.7
.9	.6004	.7508	1.3319	.7997	.1	.4	.6613	.8816	1.1343	.7501	.6
37.0	0.6018	0.7536	1.3270	0.7986	53.0	.5	.6626	.8847	1.1303	.7490	.5
.1	.6032	.7563	1.3222	.7976	.9	.6	.6639	.8878	1.1263	.7478	.4
.2	.6046	.7590	1.3175	.7965	.8	.7	.6652	.8910	1.1224	.7466	.3
.3	.6060	.7618	1.3127	.7955	.7	.8	.6665	.8941	1.1184	.7455	.2
.4	.6074	.7646	1.3079	.7944	.6	.9	.6678	.8972	1.1145	.7443	.1
.5	.6088	.7673	1.3032	.7934	.5	42.0	0.6691	0.9004	1.1106	0.7431	48.0
.6	.6101	.7701	1.2985	.7923	.4	.1	.6704	.9036	1.1067	.7420	.9
.7	.6115	.7729	1.2938	.7912	.3	.2	.6717	.9067	1.1028	.7408	.8
.8	.6129	.7757	1.2892	.7902	.2	.3	.6730	.9099	1.0990	.7396	.7
.9	.6143	.7785	1.2846	.7891	.1	.4	.6743	.9131	1.0951	.7385	.6
38.0	0.6157	0.7813	1.2799	0.7880	52.0	.5	.6756	.9163	1.0913	.7373	.5
.1	.6170	.7841	1.2753	.7869	.9	.6	.6769	.9195	1.0875	.7361	.4
.2	.6184	.7869	1.2708	.7859	.8	.7	.6782	.9228	1.0837	.7349	.3
.3	.6198	.7898	1.2662	.7848	.7	.8	.6794	.9260	1.0799	.7337	.2
.4	.6211	.7926	1.2617	.7837	.6	.9	.6807	.9293	1.0761	.7325	.1
.5	.6225	.7954	1.2572	.7826	.5	43.0	0.6820	0.9325	1.0724	0.7314	47.0
.6	.6239	.7983	1.2527	.7815	.4	.1	.6833	.9358	1.0686	.7302	.9
.7	.6252	.8012	1.2482	.7804	.3	.2	.6845	.9391	1.0649	.7290	.8
.8	.6266	.8040	1.2437	.7793	.2	.3	.6858	.9424	1.0612	.7278	.7
.9	.6280	.8069	1.2393	.7782	.1	.4	.6871	.9457	1.0575	.7266	.6
39.0	0.6293	0.8098	1.2349	0.7771	51.0	.5	.6884	.9490	1.0538	.7254	.5
.1	.6307	.8127	1.2305	.7760	.9	.6	.6896	.9523	1.0501	.7242	.4
.2	.6320	.8156	1.2261	.7749	.8	.7	.6909	.9556	1.0464	.7230	.3
.3	.6334	.8185	1.2218	.7738	.7	.8	.6921	.9590	1.0428	.7218	.2
.4	.6347	.8214	1.2174	.7727	.6	.9	.6934	.9623	1.0392	.7206	.1
.5	.6361	.8243	1.2131	.7716	.5	44.0	0.6947	0.9657	1.0355	0.7193	46.0
.6	.6374	.8273	1.2088	.7705	.4	.1	.6959	.9691	1.0319	.7181	.9
.7	.6388	.8302	1.2045	.7694	.3	.2	.6972	.9725	1.0283	.7169	.8
.8	.6401	.8332	1.2002	.7683	.2	.3	.6984	.9759	1.0247	.7157	.7
.9	.6414	.8361	1.1960	.7672	.1	.4	.6997	.9793	1.0212	.7145	.6
40.0	0.6428	0.8391	1.1918	0.7660	50.0	.5	.7009	.9827	1.0176	.7133	.5
.1	.6441	.8421	1.1875	.7649	.9	.6	.7022	.9861	1.0141	.7120	.4
.2	.6455	.8451	1.1833	.7638	.8	.7	.7034	.9896	1.0105	.7108	.3
.3	.6468	.8481	1.1792	.7627	.7	.8	.7046	.9930	1.0070	.7096	.2
.4	.6481	.8511	1.1750	.7615	49.6	.9	.7059	.9965	1.0035	.7083	.1
						45.0	0.7071	1.0000	1.0000	0.7071	45.0
Deg.	Cos	Cot	Tan	Sin	Deg.	Deg.	Cos	Cot	Tan	Sin	Deg.

APPENDIX III

Laws of Exponents

The International Symbols Committee has adopted prefixes for denoting decimal multiples of units. The National Bureau of Standards has followed the recommendations of this committee, and has adopted the following list of prefixes:

Numbers	Powers of ten	Prefixes	Symbols
1,000,000,000,000	10^{12}	tera	T
1,000,000,000	10^9	giga	G
1,000,000	10^6	mega	M
1,000	10^3	kilo	k
100	10^2	hecto	h
10	10	deka	da
.1	10^{-1}	deci	d
.01	10^{-2}	centi	c
.001	10^{-3}	milli	m
.000001	10^{-6}	micro	u
.000000001	10^{-9}	nano	n
.000000000001	10^{-12}	pico	p
.000000000000001	10^{-15}	femto	f
.000000000000000001	10^{-18}	atto	a

To multiply like (with same base) exponential quantities, add the exponents. In the language of algebra the rule is $a^m \times a^n = a^{m+n}$.

$$10^4 \times 10^2 = 10^{4+2} = 10^6$$

$$0.003 \times 825.2 = 3 \times 10^{-3} \times 8.252 \times 10^2 = 24.756 \times 10^{-1} = 2.4756$$

To divide exponential quantities, subtract the exponents. In the language of algebra the rule is

$$\frac{a^m}{a^n} = a^{m-n} \quad \text{or} \quad 10^8 \div 10^2 = 10^6$$

$$3,000 \div 0.015 = (3 \times 10^3) \div (1.5 \times 10^{-2}) = 2 \times 10^5 = 200,000$$

To raise an exponential quantity to a power, multiply the exponents. In the language of algebra $(x^m)^n = x^{mn}$.

$$(10^3)^4 = 10^{3 \times 4} = 10^{12}$$

$$2,500^2 = (2.5 \times 10^3)^2 = 2.5 \times 10^6 = 2,500,000$$

From Basic Electricity, published by the Bureau of Naval Personnel, United States Navy.

Any number (except zero) raised to the zero power is one. In the language of algebra $x^0 = 1$.

$$x^3 \div x^3 = 1$$

$$10^4 \div 10^4 = 1$$

Any base with a negative exponent is equal to 1 divided by the base with an equal positive exponent. In the language of algebra $x^{-a} = \dfrac{1}{x^a}$.

$$10^{-2} = \frac{1}{10^2} = \frac{1}{100}$$

$$5a^{-3} = \frac{5}{a^3}$$

$$(6a)^{-1} = \frac{1}{6a}$$

To raise a product to a power, raise each factor of the product to that power.

$$(2 \times 10)^2 = 2^2 \times 10^2$$

$$3,000^3 = (3 \times 10^3) = 27 \times 10^9$$

To find the nth root of an exponential quantity, divide the exponent by the index of the root. Thus, the nth root of $a^m = a^{m/n}$.

$$\sqrt{x^6} = x^{6/2} = x^3$$

$$\sqrt[3]{64 \times 10^3} = 4 \times 10 = 40$$

APPENDIX IV
Final Examination

The following questions are designed to test your overall understanding of the material presented in this book.[1]

1. State the law of attraction and repulsion of charged bodies.

2. Explain in your own words the flow of electric current in a copper wire when an electric force is applied to the wire.

3. Define in your own words the expression "difference in potential."

4. Write the equation for each of the following circuit values in terms of the other two: (a) current; (b) voltage; (c) resistance.

5. A 20-ohm, a 25-ohm, and a 100-ohm resistor are connected in parallel. What is the total equivalent resistance?

6. In a series circuit whose source voltage is 30 v., there are three resistors. The voltages across two of the resistors are 6 v. and 10 v. What is the voltage across the third resistor?

7. Refer to the schematic diagram below to answer questions (a)—(c).

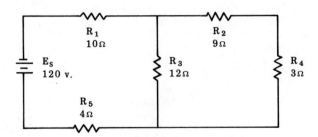

(a) What is R_t?
(b) What is the current through R_4?
(c) What is the power consumed by R_5?

8. Refer to the schematic diagram on the following page to find the answers to questions (a) and (b).

---

[1] This examination may be reproduced for personal or classroom use as needed.

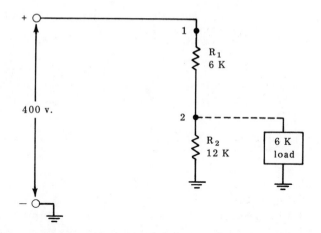

(a) What is the resistance between tap 1 and ground?
(b) What is the current through the three-kilohm load?

9. Name three ways the magnetic field strength of a current-carrying coil can be increased.

10. What is the difference between direct current and alternating current?

11. What is the effective voltage produced by an alternator if the maximum voltage is 500 v.?

12. What is the maximum voltage produced by an alternator whose effective voltage output is 500 v.?

13. What is the L/R time constant of a circuit whose total inductance is 100 mh and whose total resistance is 50Ω ?

14. What is the RC time constant of a circuit whose total capacitance is 150 microfarads and whose total resistance is 2K?

15. Refer to the schematic diagram below to answer questions (a) and (b).

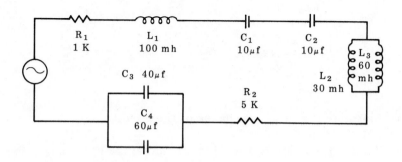

(a) Assuming that there is no mutual inductance, what is the total inductance of the circuit?
(b) What is the total capacitance of the circuit?

16. Refer to the schematic diagram below to answer questions (a)—(c).

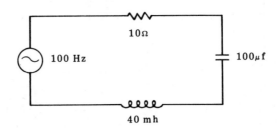

10Ω

100 Hz $100\mu f$

eee
40 mh

(a) Find X_L. (Round off to the nearest whole number.)
(b) Find X_C. (Round off to the nearest whole number.)
(c) What is the impedance of the circuit? (Use your whole numbers for X_L and X_C.)

17. In a circuit that includes inductance, capacitance, and impedance, state the phase relationship between line current and source voltage under each of the following conditions.
    (a) The circuit is at resonance.
    (b) X_L is greater than X_C.
    (c) X_C is greater than X_L.

18. In a series L—C—R circuit, under what specific condition is impedance minimum and line current maximum?

19. What is average, or true, power?

20. Refer to the schematic diagram below to answer questions (a)—(e).

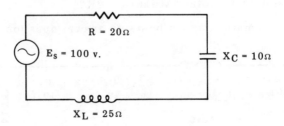

$R = 20\Omega$

$E_s = 100$ v. $X_C = 10\Omega$

eee
$X_L = 25\Omega$

(a) Calculate the total impedance of the circuit.
(b) Calculate the line current.
(c) What is the phase relationship, including the phase angle, between the line current and the source voltage?
(d) What is the apparent power of the circuit?
(e) What is the average power of the circuit?

Final Examination Answers

If your answers to the test questions do not agree with those that follow, review the chapters indicated in parentheses after each answer. Detailed solutions to problems are given where appropriate.

1. Like charges repel each other and unlike charges attract each other. (1)

2. When an electric force is applied to a copper wire, free electrons are displaced from the copper atoms and move along the wire, producing electric current. (1)

3. The difference in potential, or electromotive force, is the force that causes free electrons to move in a conductor as electric current. (2)

4. (a) $I = \dfrac{E}{R}$ (3)

   (b) $E = IR$ (3)

   (c) $R = \dfrac{E}{I}$ (3)

5. 10Ω (4) Solution: $\dfrac{1}{R_t} = \dfrac{1}{R_1} + \dfrac{1}{R_2} + \dfrac{1}{R_3}$

   $$= \dfrac{1}{20} + \dfrac{1}{25} + \dfrac{1}{100} = 0.05 + 0.04 + 0.01$$

   $$= 0.1$$

   $$R_t = \dfrac{1}{0.1} = 10$$

6. 14 v. (4) Solution: $E_S - E_1 - E_2 - E_3 = 0$

   $$E_S - E_1 - E_2 = E_3$$

   $$30 - 6 - 10 = 14$$

7. (a) $R_t = 20\Omega$ (4, 5) Solution: To find R_t, first find the resistance of the parallel branches. R_3 forms one branch, while R_2 and R_4 form the other branch. To find the equivalent resistance of R_3, R_2, and R_4, use the "product over sum" method.

   $$\dfrac{12 \times 12}{12 + 12} = \dfrac{144}{24} = 6\Omega$$

   Then add the series resistance, R_1 and R_5, to the equivalent parallel resistance of 6Ω. $R_t = 10 + 4 + 6 = 20\Omega$

   (b) The current through R_4 is 3 a. (4,5) Solution: First find I_t.

   $$I_t = \dfrac{E_s}{R_t} = \dfrac{120 \text{ v.}}{20\Omega} = 6 \text{ a.}$$

Find the voltage across R_1 and R_5.

$$E_1 = I_t R_1 = 6 \times 10 = 60 \text{ v.}$$

$$E_5 = I_t R_5 = 6 \times 4 = 24 \text{ v.}$$

Since the total voltage across R_1 and R_5 is 60 v. + 24 v., or 84 v., the voltage across each parallel branch is 120 v. − 84 v., or 36 v.

The current through the parallel branch that includes R_4 is found as follows.

$$I = \frac{E}{R} = \frac{36}{9 + 3} = \frac{36}{12} = 3 \text{ a.}$$

Since the same current flows through each resistor in that branch, the current through R_4 is 3 a.

(c) The power consumed by R_5 is 144 w. (4, 5) Solution: Since total current (6 a.) flows through R_5,

$$P_5 = I^2 R = 6^2 \text{ a.} \times 4\Omega = 36 \text{ a.} \times 4\Omega = 144 \text{ w.}$$

8. (a) 10 K (5) Solution: The equivalent resistance of the parallel combination of R_2 and the load is 4 K. $R_t = 6 \text{ K} + 4 \text{ K} = 10 \text{ K}$

(b) 0.0266 a. (5) Solution: First find I_t, which is 400 v. ÷ 10 K = 0.04 a. Find the voltage drop across R_1, which is 0.04 a. x 6 K = 240 v. The remaining voltage, 160 v., is across the 6 K load. The current through the load is 160 v. ÷ 6 K = 0.0266

9. Increase the number of turns of wire in the coil, increase the current flow through the coil, and select a core material with higher permeability. (6)

10. Direct current flows in only one direction, while alternating current periodically reverses the direction of flow. (7)

11. 353.5 v. (7) Solution: Effective voltage is equal to maximum voltage multiplied by 0.707. 500 x 0.707 = 353.5

12. 707 v. (7) Solution: Maximum voltage is equal to effective voltage multiplied by 1.414. 500 x 1.414 = 707

13. 2 milliseconds (8) Solution: Divide 100 mh by 50 Ω. Since inductance is in millihenries, the time constant is in milliseconds.

14. 300 milliseconds (9) Solution: Multiply 150 microfarads by 2K. The time constant is 0.3 second or 300 milliseconds.

15. (a) 120 mh (8) Solution: Series and parallel inductances are calculated in the same way as resistances.

(b) 105 μf (9) Solution: Capacitances in series are calculated in the same way as resistances in parallel; and capacitances in parallel, as resistances in series.

16. (a) $X_L = 25\Omega$ (10)

   $X_L = 2\pi fL = 6.28 \times 100 \times 0.040 = 25.12$

   (b) $X_C = 16\Omega$ (10)

   $$X_C = \frac{1}{2\pi fC} = \frac{1}{6.28 \times 100 \times 0.0001} = 15.92$$

   (c) $Z = 13.45\Omega$ (10)

   $$Z = \sqrt{R^2 + (X_L - X_C)^2} = \sqrt{10^2 + (25 - 16)^2} = \sqrt{100 + 81}$$
   $$= \sqrt{181} = 13.45$$

17. (a) Line current and source voltage are in phase. (At resonance, $X_L = X_C$.)   (11)
   (b) Line current lags source voltage.   (11)
   (c) Line current leads source voltage.   (11)

18. When the circuit is at resonance; that is, when $X_L = X_C$.   (11)

19. Power that is dissipated as heat.   (11)

20. (a) $Z_t = 25\Omega$ (11)   Solution: $Z = \sqrt{R^2 + (X_L - X_C)^2}$

   $$= \sqrt{20^2 + (25 - 10)^2} = \sqrt{20^2 + 15^2}$$
   $$= \sqrt{400 + 225} = \sqrt{625} = 25$$

   (b) The line current is 4 a.   (11) $I_t = \dfrac{E_s}{Z_t} = \dfrac{100}{25} = 4$

   (c) Line current lags source voltage by 36.9 degrees.   (11) The circuit is inductive, because X_L is greater than X_C. In the impedance triangle, R is the adjacent side, and X is the opposite side. Tan θ is equal to the opposite side divided by the adjacent side. $15 \div 20 = 0.75$. Look up tan 0.75 in the tables, Appendix II. Tan 0.7508 is 36.9 degrees.

   (d) The apparent power of the circuit is 400 w.   (11) $P_{ap} = E_s I_t = 100 \times 4 = 400$

   (e) The average power of the circuit is 320 w.   (11)   Average power is the power consumed by resistance, since power apparently consumed by inductors and capacitors is not dissipated as heat but returned to the source. To find the power consumed by the resistor,

   $$P = I^2 R = 4^2 \times 20 = 16 \times 20 = 320.$$

   Or $P_{av} = EI \cos \theta$. The cosine of 36.9 degrees, the phase angle, is 0.8. (The tables in Appendix II show 0.7997.) $P_{av} = 100 \times 4 \times 0.8 = 320$

Afterword

Now that you have learned the basic principles of electricity, it's on to the world of electronics if that is where you wish to go. You have laid the groundwork; you will find it relatively easy to move to the next plateau—electronics.

But what is the difference between electricity and electronics? The old definition is obsolete. We used to say that if the circuit had a vacuum tube, it was electronic. But how often do you see a vacuum tube? It does exist for special applications, but you can work a year with electronic equipment and never see one.

Modern electronics is the world of the transistor, the integrated circuit, and the microprocessor. It is the world of digital circuits, the application of logic on printed-circuit boards. And it is a world that is simpler than you might expect.

Today, if you wish to work in the field of electronics, it is not enough to study electronics. You must also study digital electronics, which is the special application of basic electronic principles that makes the computer possible.

Any bookstore or library carries a great many books on basic electronics. Perhaps too many. But if you feel comfortable with the self-teaching method of study you followed in this book, I can recommend two fine texts. They are Electronics, by Harry Kybett, and Digital Electronics, by Harry Kybett with Vaughn D. Martin. Both are Self-Teaching Guides from John Wiley & Sons.

If you studied Basic Electricity simply to increase your store of knowledge, I congratulate you. You have done it. But if you want to open the door to the high-technology world of electronics, you have the key. Use it!

Index

NOTES

NOTES

NOTES

NOTES

NOTES